A

# EXTEL 100

# EXTEL 100

## THE CENTENARY HISTORY OF THE EXCHANGE TELEGRAPH COMPANY

### J. M. SCOTT

INFORM

LONDON/ERNEST BENN LIMITED

1972

First published 1972 by Ernest Benn Limited
Bouverie House, Fleet Street, London, EC4A 2DL

Distributed in Canada by
The General Publishing Company Limited, Toronto

© The Exchange Telegraph Company (Holdings) Limited 1972

Printed in Great Britain by Wightman & Co. Ltd.
a Member of the Extel Group

ISBN 0 510-27951-1

# Contents

# List of Illustrations

# *Acknowledgements*

The author and the publishers wish to record their grateful thanks to copyright owners for the use of the illustrations listed below:

Baron Studios, for 48

Tomas Jaski Ltd., for 15, 16, 18, 23, 26, 32, 39, 44, 49

Hugh Sibley, for 30

Press Association, for 19, 20

Crown Copyright, Science Museum, London, for 24

British Rail, London Midland Region, for 45

P.I.C. Photos Ltd., for 40

Syndication International, for 21, 22

Mack of Manchester Ltd., for 17

Illustrated London News, for 7

Sport and General Press Agency, for 36

CHAPTER I

# A Combination of Circumstances

RARELY if ever can one pin-point when something began. A human creature has parents – who must first meet. So it is with inventions, organizations. Each depends, like the living creature, on an idea and a combination of circumstances without which it could never take shape.

The birthday which this book celebrates is 28 March 1872. But we must go back a score of years before that to rally and review the diverse circumstances which combined to bring that birth about. Let us start at the Great Exhibition of 1851.

Exhibitions were nothing new, but this was the first truly international one. The British Empire, then at the zenith of its wealth and self-confidence, was glad to display its products – minerals and raw materials, machinery, manufactures, and fine arts – alongside those of other nations.

The section in the Crystal Palace which most drew the crowds was the mechanical. In less than a generation steam had conquered this country as the motor-car has not quite managed to do in over seventy years. The face of Britain had been changed by railways, tunnels, long bridges. Locomotives could already cover a mile in a minute – a revolutionary achievement to people who until so recently had had to depend upon the horse or their own legs to get about.

True, steam had not yet conquered the sea. The great clipper races were still to come – in the 1860s – and the reign of the wind ships in fact lasted until the opening of the Suez Canal. Even then they remained in some ways preferable to steam-ships, for however long the voyage they did not have to put in to refuel. But steam-cum-sail ships were increasingly used. Steam was the coming thing. It was potentially capable of any

1

speed, any power. Nature had been put in harness by man.

There were giants in those days. When one considers that memory of our present political leaders may not last a hundred years one is impressed by the still evocative names of Victorian Prime Ministers – Peel, Palmerston, Gladstone, Disraeli. In many other walks of life there were outstanding, still-remembered figures. Here we need mention only three. The first is Joseph Paxton, the farmer's son who became head gardener to the Duke of Devonshire and designed the Crystal Palace, originally as a doodle on blotting-paper. The second is Isambard Kingdom Brunel, the five feet four inches 'Little Giant' whose engineering creations were always revolutionary and generally out-size, who built twenty-five railways, numerous bridges, also dry docks, enormous steamships and, at the age of twenty, the Thames Tunnel which was the first shaft under any river in the world. The third is Professor William Thomson, who became Lord Kelvin, the greatest physicist of his time. These three men typify the spirit of the Victorian age.

Although most techniques were improving rapidly, that concerned with the transmission of messages was a notable exception. Since early in the century newspapers had printed overseas dispatches 'Received by Telegraph'. But this only meant that they had been transmitted from the port to the capital along one of the chains of semaphore apparatus first installed during the Napoleonic war. Not until 1850 were they 'Received by Electric Telegraph' – by land line. And still they only came by this means from the port, for apart from the Dover-Calais cable laid in 1851 there was no cable across the sea. Thus a message sent from Southampton to New York took as long as the ship which carried it – ten days or more, dependent on favourable winds and fair weather. There had never been such a difference between land and sea communications. To right this, the problem of insulation in water had to be solved.

One of the most fascinating aspects of human development is how man has discovered the potential uses of natural products. Gutta-percha had for centuries been used by the Malays for whips. In 1843 Dr William Montgomerie of the Indian Medical Service used it for the handles of surgical instruments. As such it was introduced to England; but not until some years later was it discovered that gutta-percha, very hard yet pliant, was entirely non-conductive of electricity. Thereafter it was used as an insulator, particularly under water, until rubber supplanted it.

When England and France had achieved telegraphic communication the next great venture had to be the laying of an Atlantic cable. In 1854 an English engineer named Frederick Gisborne went to the United States and persuaded the American financier and former paper merchant, Cyrus Field, to form an Atlantic cable company. Americans in general were enthusiastic about the idea, but their enthusiasm stopped short of risking their money. Alone among his countrymen, Field bought eighty-eight £1,000 shares in the company he had formed. British financiers bought two hundred and twenty-nine. Brunel was deeply interested in the project and sent copies of memoranda to Field about cable-laying.

The first plan was for two ships to meet in mid-Atlantic, share out the cable, and sail away from each other to their respective shores. In 1857 HMS *Agamemnon* and the U.S. warship *Niagara* met halfway. But the cable which they shared was not nearly strong enough, and parted after only a couple of miles had been paid out.

In 1858 the same two ships started laying cable from their respective shores, with the intention of joining the ends when they met. The cable broke six times, but was repaired. The *Agamemnon* and *Niagara* met. The junction was made. Four hundred messages were passed across the Atlantic. Then the signals faded out, and the line went dead. There was a good deal of argument as to why this had happened, but in fact so strong a current had been used that the insulation had failed.

Field had lost £$\frac{1}{2}$ million. But he was not discouraged. He merely waited for a favourable opportunity to try again.

To explain how this opportunity arose we must go back again to 1851. Inspired by the Great Exhibition, Brunel then had the greatest of all his ideas – in sheer size. This was for a ship far bigger than any that had been dreamed of before.

The ship (as yet she had no name) took six years to build and three months to launch into the Thames from her wooden cradle on the muddy shore of the Isle of Dogs. She had by that time cost £800,000 and seven human lives. She was no ordinary ship. She was six hundred and ninety-two feet long, with a beam of one hundred and twenty feet and a displacement of 22,500 tons. She was designed to carry 4,000 passengers. She was five times larger than any other vessel at that time and she held the record for tonnage until the *Lusitania* was launched in 1906.

She had five funnels and six masts which carried 6,500 square yards of sail – well over an acre of canvas. The masts, being too many to be designated in nautical terms, were called after the days of the week, Monday to Saturday. The saloon was more spacious than any other, even to the present day. One of the funnels, which passed through it, was entirely encased in mirrors, adding to the impression of size. Her plates were of iron almost an inch thick: she had a double hull with a three-foot space between. During her building one of the thousands of riveters went missing and was believed to have fallen into this space, his cries for help being drowned by the continuous hammering required to drive 3 million rivets by sledge-hammer. This man's ghost haunted the ship all her life. It was often heard hammering and was held responsible for the almost uncanny bad luck which dogged the ship. In fact, this is one of the very few ghost stories which have a satisfying end, for when the vessel was broken up in the last decade of the century the riveter's skeleton was found between the hulls.

Professor John Scott Russell was responsible for constructing the hull and the paddle engines, and the James Watt Company for the screw engine – for she had a twenty-four-foot propeller besides her fifty-eight-foot paddles. When at last they got her into the water it took another eighteen months to fit her out.

She had been named *Leviathan* at the beginning of the first launching attempt, when she stuck in the mud. But long before she was in the water she was being called *Great Eastern*; and the name stuck until it became official. This derived from the Eastern Steam Navigation Company which financed her building. But she broke that company before she put to sea, and it was the Great Ship Company that sent her on her maiden voyage from the Thames to Holyhead on 12 September 1859.

She started in grand style, doing twelve knots in the Estuary. But off Hastings she blew out a funnel, killed six men, and shattered all the mirrors in the saloon.

That was the beginning of seven years' bad luck.

She did not make her first voyage to New York until June 1860. Constant delays had so much discouraged those who might have sailed in her that she only carried thirty-five paying passengers instead of the 4,000 for which she had been designed. Since her crew was four hundred and eighteen, the voyage was made at a heavy loss. But she got a good welcome from the New World. The First Officer on that voyage was W. H. Davies, a name which will be heard again.

In 1862, the *Great Eastern* made in ten days her first financially successful Atlantic crossing, with 1,530 passengers and a valuable cargo. Off Long Island her captain, Walter Paton, took on a pilot for the Sound. The ship had to pass through a channel which according to the chart was nowhere shallower than thirty-five feet. Since the *Great Eastern*, even with her exceptionally heavy load, drew no more than thirty feet the pilot took her through under both screw and paddles. She went over an uncharted rock which made a gash in her bottom nine feet wide and eighty-three feet long – and there was no dry dock large enough to take her.

She was brilliantly repaired in the water, in spite of a strike caused by the phantom riveter. But the bill was £70,000. One more unsuccessful voyage put the company so deep in the red that the *Great Eastern* was put up for auction, and knocked down for £25,000 – just over a £1 a ton. But this scrapyard purchase proved to be the start of the ship's useful life, for the buyer was Daniel Gooch.

Gooch was a remarkable man. As an engine-driver he had taken Queen Victoria on her first railway journey – while the nation prayed for her safety. As Brunel's right-hand man he rose rapidly from his humble beginning to become a Member of Parliament with the unique distinction of never once speaking in the House. He knew that Cyrus Field was interested in the *Great Eastern,* and also that he was short of cash. He told Field that he could have the ship. If she failed to lay the cable there would be no charge whatever: if she succeeded he would like

£50,000 of cable stock. Field jumped gratefully at this offer, and made Gooch and his partner directors of the Cable Construction Company.

Field's interest in the *Great Eastern* was not only the sympathetic one that he had lost £½ million and she £1 million to the companies who had owned her. The necessity of using two ships had complicated the former cable-laying attempts. And here was a vessel – the only one in the world – which could carry 2,300 miles of cable in her holds. But the huge size of the *Great Eastern* created its own danger. The cable (a section is preserved in the house of Douglas Anderson, great-grandson of the Captain) was the thickness of a cricket stump. The *Great Eastern* had a beam of forty yards and was well over two hundred long. The cable would inevitably suffer stress and strain as it was paid out into the ocean. The slightest mishandling of the huge vessel would cause it to snap. Therefore the skill and temperament of the captain would be of vital importance.

The right man by reputation was Captain James Anderson. He commanded the *China* for the Cunard Company. But in the friendly, sporting spirit which characterized the whole affair, Cunard willingly gave Anderson leave. He saw to the considerable alterations which were necessary on the *Great Eastern*: the saloons, cabins, and holds were replaced by three huge cable drums; a funnel and two boilers also made room for wire. In May 1865 she began taking on board her strange cargo at Sheerness, and in July she sailed for Foilhummerum Bay on Valentia Island in Ireland's County Kerry where a great crowd of people with plenty of poteen collected to do her honour. There the bight of the cable was slipped from the *Caroline,* which had carried it ashore, and the *Great Eastern* sailed north-west.

> The brake [of the cable] was eased, and as the *Great Eastern* moved ahead the machinery of the paying-out apparatus began to work, drums rolled, wheels whirled, and out spun the black line of the cable, and dropped in a graceful curve into the sea over the stern wheel.

The quotation is from the despatches of W. H. ('Bull Run') Russell, the famous war correspondent of *The Times* who was on board together with Daniel Gooch, Cyrus Field, Professor William Thomson, and teams of technicians, American and British.

A full account of this remarkable voyage and that which followed it may be found in James Dugan's *The Great Iron Ship* (Hamish Hamilton, London). Here they can be little more than summarized. The *Great Eastern* steamed away from Ireland at six knots. As the cable was paid out over the stern, the technicians on board sent a continuous stream of signals through it to the electricians on land. On the ship a perpetual watch was also kept on the astatic mirror galvanometer, brain child of Professor Thomson. This measured the ohm resistance of the cable, throwing a tiny point of light on a graduated scale so long as the cable was alive. If the light jumped off the scale it meant a fault or a break. Then a gong was to be struck.

The *Great Eastern* had covered eighty-four miles when the gong boomed. Captain Anderson immediately stopped the ship. It was then necessary to pick up the line until the fault was located. To do this the cable had to be cut and passed to the bow, where it could be wound in at one mile an hour on a steam windlass (the ship having been turned about) without strain on the cable.

After ten hours of winding the fault was found – a small extraneous piece of wire had pierced the insulation and allowed the current to escape. It was difficult not to suspect sabotage. Certainly Russell did.

The fault was cut out, the free end spliced to the drum supply, the cable passed back outboard of stays and other obstructions throughout the furlong of deck, and the ship swung back upon her course again. . . . A moment's thought will suggest some of the difficulties involved.

There were other scares, some which righted themselves mysteriously but satisfactorily, some which involved cutting the cable, back-tracking, re-winding, eliminating the fault, and finally splicing the cable again. On one occasion Captain Anderson spent twenty-six hours delicately manoeuvring the enormous ship to avoid damaging the spider's thread of cable.

When they had covered 1,186 miles there was yet another apparent fault and check to the advance. Professor Thomson said modestly that he thought the trouble would work itself out, and explained why in a few technical sentences. But he was not yet Lord Kelvin, and he was overruled by another of the experts on board. The cable was passed to the bow, and rewinding began. Then for some reason unknown it suddenly parted, flew through the stoppers before the brakeman could halt it, flicked like a great whip through the air and lashed into the ocean, making a momentary weal on the smooth surface, then sinking without further trace.

They did not then know the depth – which was later found to be 2,000 fathoms. The cable, just over an inch in diameter, lay more than two miles down at the bottom of the sea. 'Cable?' said Captain Anderson, 'Thread!' But he decided to fish for it. He steamed up-wind, turned about and sailed – literally sailed for the first and only time – down-wind, trailing a grapnel. After fishing all night they hooked something and brought it half way to the surface before the grapnel line parted. Captain Anderson marked the place with a flagged buoy held to a mushroom anchor by three miles of unused cable. And they tried again. They fished for nine foggy days and nights, twice more hooking the cable (presumably) and bringing it some way towards the surface before the grapnel line parted, Captain Anderson finding and re-finding his marker buoy on each occasion by what seemed instinct rather than navigation. But when the grapnel line parted for the fourth time, and there was no more rope, the *Great Eastern* could only turn for home.

Yet no one concerned was discouraged to the point of giving up – far from it. By the next spring another £1 million of capital had been raised and another 3,000 miles or so of cable made. The cable had consisted of seven copper wires twisted in rubber, this wrapped in eight thicknesses of rubber compound, and the whole covered by ten charcoal-iron wires spun in tarred manilla. An intricate type of cable, and the new one was thicker and still more complex. Yet this huge length of it was turned out in little over six months. It says something for British engineering a century ago.

The 1866 voyage of the *Great Eastern* was almost dull because it was so successful. On 26 July the cable was carried ashore on Newfoundland at a place called Hearts Content. It worked, and went on working. Congratulatory messages sizzled to and fro at 5s a word – and the latest stock market quotations of London and New York were exchanged. Then the ship which had never before done anything but lose fortunes picked up the broken cable of the previous year, buoyed it to make a second transatlantic string later, and sailed on to England where her captain and five others concerned with the enterprise were knighted by Queen Victoria.

Captain Sir James Anderson made one more voyage in command of the *Great Eastern*. It was like a dream of her youth: it was scarcely more controlled by reason. Louis Napoleon, who proclaimed himself the third Emperor, set his heart on a Paris exhibition to outshine all exhibitions in 1867. But how could the vast army of American millionaires be brought to it? '*Le Grand Oriental!*' he cried in a moment of inspiration.

Daniel Gooch was no fool. The *Great Eastern* was out of a job. Louis Napoleon could have the ship, but he was going to pay handsomely. Gooch and his fellow directors thought up one of the best rumours there have ever been. It became known in Paris that His Transcendent Highness, Sultan Abdul-Aziz of Turkey, meant to have the *Great Eastern* as a harem, and nothing would stop him. Louis Napoleon stopped him by chartering the ship for £80,000.

Out came all the cable gear and in went the saloons and staterooms with even more gilded luxury than Brunel had originally intended. Naturally this cost a lot of money – but who cared?

The first voyage westward began on 26 March. The prime intention was to pick up the golden cargo of American millionaires, but there were rich and famous personages among the passengers of this outward voyage. One – though as yet scarcely rich or famous – was Jules Verne. To him we owe a few references to Captain Sir James Anderson. He did card tricks at a passengers' entertainment. When a cyclone struck the ship he smiled as usual and refused to alter course. 'I was astonished at the Captain's obstinacy,' wrote Verne.

The *Great Eastern's* stay in New York was very expensive.

Harbour dues, pilotage, agents' commissions and a number of etceteras came to a couple of thousand pounds. Only one hundred and ninety-one passengers, instead of the ten or twenty times this number which had been expected, booked for the voyage to France. Worse still, when the ship returned to Europe, proving that Louis Napoleon's plan was a flop, the French, who had run up enormous bills for refitting the *Grand Oriental*, refused to meet most of them. Captain Anderson had the unpleasant duty of telling his crew that the Company were unable to pay them.

The *Great Eastern* wriggled out of her troubles, more or less, as she always had before. She went back to cable-laying. She spun three more threads across the Atlantic. But not under the command of Captain Anderson. He had had enough of that side of the business. He cannot have failed to notice how many of the first transatlantic messages had been concerned with the price of stocks and shares. London was now linked not only with New York but with all the capitals in Europe. Anderson had become a close friend of Field, who had always believed there was money to be made out of instant information about money. And one of the people most interested in the new cables was Paul Julius Reuter, founder of the first news agency.

Anderson had made a name for himself. He had a good brain and good health. Recent years had given him some business experience, and he had met a number of influential people . . .

All the essential circumstances have been combined, and we have almost reached 1872. Sir James Anderson, the sailor turned businessman, held a number of meetings in Birch's restaurant near the Stock Exchange. The result of these will be described in the next chapter.

The reader has been given one aspect of the period. But it seems worth while to take a closer look at the birthday year. At least it may be interesting to see how different – or how similar – everday life was then and now.

In 1872 the Channel Tunnel Company was registered. There were violent riots between Protestants and Catholics in Northern Ireland. The National Revenue was £76·6 million.

The Bank Rate rose to nine per cent., and there was near-panic in the City. Income tax was reduced from 6d to 4d in the pound. Disraeli stated that Gladstone's Ministry 'harassed every trade, worried every profession and assailed or menaced every class, institution and species of property in the country'. Mr Biggar caused reporters to be excluded from debates in the Commons; much discussion ensued; finally Disraeli's resolution that strangers were not to withdraw without a vote of the House or order of the Speaker was adopted.

The Meteorological Office began issuing daily weather charts. Girton College, 'for the higher education of women,' was opened in Cambridge. (In 1972 several of the men's colleges are admitting women undergraduates for the first time.) Builders struck for a nine-hour day and 9d an hour. A meeting at the Mansion House advocated the introduction of the metric system.

In the United States a proclamation was issued against the Ku Klux Klan. (Negro equality with whites had been recognized two years before.) Sir Bartle Frere went to Zanzibar on a mission to suppress the slave trade. Dr Livingstone, who had been found by Stanley the year before, was dying at Ujiji, and his letter about the slave trade was published in *The Times*.

At the Church Congress it was stated:

> We live in an age when all opinions and beliefs are keenly criticized and when there is less inclination than ever before to respect authority in matters of opinion. In every state, in every religious community, almost in every family, the effect of this unsettled condition may be traced.

The Crystal Palace, by this time re-erected at Sydenham, was used for a thanksgiving festival for the recovery of the Prince of Wales, for a bird show, a cat show, a dog show, and the Conservative Conference.

There was a Post Office scandal: money from other departments had been spent on the telegraphic service without authority of Parliament. During the year 15½ million telegrams were sent and 885 million letters. Letters cost 1d for one ounce and 1½d for two. The pigeon post between London and Tours, started during the Siege of Paris of 1870, was abandoned.

The population of England, Scotland and Wales was twenty-five million, less two hundred and fifty-seven people who were murdered during the year. This was a higher annual rate than has since been achieved so far.

# CHAPTER 2

# The First Ten Years

I hereby certify that The Exchange Telegraph Company, Limited, is this day incorporated under the Companies Act, 1862, and that this Company is Limited.

Given under my hand at London this twenty-eighth day of March, One thousand eight hundred and seventy-two.

E. C. CURZON
*Registrar of Joint Stock Companies*

THE FOUNDERS of the company, as stated in the Articles of Association, were Sir James Anderson and George Baker Field. The capital was £200,000, divided into 6,000 'A' shares bearing interest at ten per cent and 14,000 'B' shares to be allotted as fully paid shares. Following the custom of the times the 'A' shares were not fully paid up, so the Directors were able to make calls on the shareholders when money was short during the early critical years. The subscribers to the Memorandum of Association, each of whom had taken one hundred shares, were Lord Borthwick, who was a stockbroker, Sir James Anderson, Edward Fowler Satterthwaite, a stockbroker, George B. Field, landed proprietor, Claudius Edward Habicht, a banker, Julius Beer, a merchant, and Charles Smith Seyton, who was described as a gentleman of no occupation. The objects of the company were set out as follows:

1. To erect, maintain, and work between any office of the Company, or any Stock or Commercial Exchange, in any city or town in the United Kingdom, and the offices of the same city or town of any Broker, Share-dealer, Banker, or Merchant, or other person having business with or upon any such Stock or Commercial Exchange, wire and apparatus, for the purpose of transmitting to such persons

13

respectively, quotations of the prices of stocks or shares, and produce, arrivals or departures of shipping, or other business intelligence.

2. To erect, maintain, and work telegraphs, and transmit telegrams, in accordance with the terms of any licence granted by Her Majesty's Postmaster-General, or other competent authority.

3. The doing of all such things as are incidental or conducive to the attainment of the above objects.

The interests of the Company, it will be noted, were exclusively financial and commercial.

At their first meeting on 12 April 1872, the subscribers voted themselves Directors. But they took some time to find a permanent Chairman, Julius Beer, Sir James Anderson, Lord Borthwick and C. S. Seyton all taking the Chair until Lord William Montagu Hay was elected by the Directors in time to preside at the first General Meeting on 25 July.

Quite a lot had been achieved in these first four months. There had been eight Directors' meetings. The first seven were held in temporary premises at 11 Old Broad Street, E.C.2, but by mid-July a fourteen years' lease had been taken of 17 and 18 Cornhill for £250 a year. This was to remain the Exchange Telegraph's home until 1919 when the freeholders, Lloyd's Bank, 'required' the premises. W. H. and G. C. Bompas had been selected as the Company's solicitors and Captain W. H. Davies as acting Secretary 'at the rate of £300 p.a. . . . until one hundred instruments are at work when his salary shall be £500 p.a.' – a productivity agreement long before the term was invented.

The Cornhill premises had been fitted with batteries, wires and other apparatus sufficient for two hundred instruments. These early tape machines were being imported from the United States, this being the only work supervised by George Field before he sold his shares and faded from the scene. Negotiations for the necessary patent rights were under way. A licence had been received from the Postmaster-General authorizing the Company's system of telegraphy to be carried out in London within a radius of nine hundred yards of the Stock Exchange, as also in Liverpool, Manchester, Leeds, Birmingham, Edinburgh, Glasgow, and Dublin; but arrangements with the Stock Exchange had not yet been completed. Messrs. Glyn & Co. had been appointed the Company's bankers – Williams & Glyn's Bank Limited are to-day's bankers. The financial position of the Company as submitted at this first General Meeting was as follows:

| | | | | | | |
|---|---|---|---|---|---|---|
| Cash by first call on 'A' shares | | | | £3,000 | 0 | 0 |
| Expenditure to date | £497 | 8 | 4 | | | |
| Bills received for payment | £916 | 14 | 4 | £1,414 | 2 | 8 |
| | | | | | | |
| Balance at Bankers after payment | | | | £1,585 | 17 | 4 |
| | | | | £3,000 | 0 | 0 |
| Estimated amount of outstanding accounts | | | | £444 | 0 | 0 |

The first full financial year, July 1872 to July 1873, was largely occupied in negotiation with Managers of the Stock Exchange for permission to have access to the House. The Directors of the Exchange Telegraph showed great persistence and a sugaring of guile. The Managers of the Stock Exchange did not at first look favourably on any invasion of their privacy. The Directors, therefore, were careful that no letter addressed to the Managers should offer an opportunity for a direct and definite 'no' which would have closed the question. Meanwhile they brought personal pressure to bear on individual Managers (one smells some expensive entertaining between the lines of the minutes), they canvassed members of the Stock Exchange for signatures to an appeal to the Managers that the Exchange Telegraph's service should be applied for, they sold this service to outside firms for £50 a year per instrument and supplied them by means of transmissions from the office of Messrs. Borthwick and Wark (both of whom were Directors) in Bartholomew House, then in Bartholomew Lane. This service consisted originally only of the supply of stocks and shares prices by direct wire to a type of tape machine in each subscriber's office.

When four hundred and six signatures of Stock Exchange members had been obtained, and when the Managers were reported to be in an amenable mood, a direct request was made; and even then the outside subscribers were cut off for a month to avoid any risk of the Managers being offended. This careful handling resulted in permission to place an operator in the Settling Room, and to furnish instruments to non-members.

The numbers of subscribers immediately rose to ninety-seven – three short of the target given to the Secretary, W. H. Davies. He none the less obtained his reward by being given a place on the Board and made Managing Director at a salary of £500 a year.

Two other appointments are of interest. George Christie was made the Company's reporter at £600 a year, out of which salary he was to obtain and pay two clerks to help in collecting financial news. More important was the result of terminating the contract of the chief engineer, an American named C. A. Callahan who had been telegraphed for by G. B. Field after the first Board Meeting, and who had been responsible for the purchase and installation of the American instruments which were exclusively used for the first year, and to a lesser extent thereafter. Neither he nor the instruments appear to have been entirely satisfactory. Exactly how he failed is not clear, but it was found that equally good British instruments could be bought more cheaply. As a result of Callahan going, the assistant engineer was promoted chief at £300 a year. This was Frederick Higgins, something of a mechanical genius who by good fortune had been recruited young, and who was to hold the post with great distinction for many years.

Captain W. H. Davies, first as acting Secretary and then as Managing Director, was kept busy obtaining wayleaves for the passage of wires over buildings in the vicinity of the Stock Exchange. In the late summer of 1872 overhead men were running wires between Bartholomew House, later to become the telephone exchange for stockbroker subscribers, and Austin Friars, and permission was being sought to carry cable over the Sun Fire office and the Royal Exchange.

Obtaining wayleaves was a major task, for the Company of course had no powers to force landlords to give permission. Every agreement had to be arrived at individually by persuasion alone; and if in a single instance this was not forthcoming the line had to be re-routed. It was like a game of snakes and ladders. But on 30 November Stock Exchange quotations began to be sent direct from the Settling Room to the Company's offices in Cornhill, and from there transmitted to subscribers.

Negotiations were also started with Lloyds to supply insurance agents with news relating to casualties and losses, the suggested fee being £30 a year.

Construction costs were naturally heavy – over £8,000 by July 1873 – and the Company was not yet making a profit. But some £1,500 had by then been received from the rent of instruments in London. Inquiries had come in from brokers as far afield as Glasgow, and the Board confidently looked forward to 'a very satisfactory net revenue' before the end of 1873.

The minutes of Board meetings in these early years are tantalizingly brief – partly, perhaps, because everything had to be recorded in long-hand – and there remains no correspondence to fill in the gaps. But interesting points may be deduced. Charles Seyton, the only member of the Board described as a gentleman of no occupation, was always busy. Towards the end of 1873 he was in the United States, examining new instruments and reporting on the possibilities of supplying New York Stock Exchange prices. Almost every Board meeting considered some point he had raised by letter or telegram. His commendable energy and drive were balanced by the carefulness with which he put his hand in his own pocket. The Directors discussed an expenses claim of his for £200. And a note in the account book shows that he refused during his absence from England to pay the quarter's rent on the instrument he had privately hired – £12 10s 0d.

The Company's character also comes out. They gave Seyton all he asked. And to Frederick Higgins, after less than a year as chief engineer, they gave an honorarium of a year's salary and a rise of £50 a year for his excellent service.

As busy as anyone was Captain Davies. He was by training a sailor, not a businessman, but his fellow Directors referred every question to him, in his capacity of Managing Director, for examination and report – the possibility of placing instruments in newspaper offices, further negotiations with Lloyds, the appointment of agents in Manchester and Liverpool, the practicability of the Glasgow fire alarm service. But it was all routine – until Wednesday 14 January 1874 when an emergency Board Meeting was called at two hours' notice 'for the purpose only of considering the best steps to pursue in regard to the recovery of the Bonds taken from the Company by Laurence Archdeacon on the night of Tuesday 13 inst.'.

No other business was discussed. It was resolved 'that every possible means shall be employed for the apprehension of the said Laurence Archdeacon and the recovery of the stolen property.' That is all the minutes record at the time, and they scarcely mention the matter thereafter. But a search of contemporary newspapers reveals a human story not without touches of humour.

Archdeacon, who at the time of the meeting was only a suspect although the minutes labelled him thief, was an accountant who had been employed for eighteen months, virtually from the inception of the Company, at £1 a week. He had a reputation for diligence and sobriety. But on Saturday 10 January he took a drink or two, then more. He liked the result and decided that at thirty-three he had wasted his years being sober and diligent.

This came out later. At the time nobody knew of his weekend escapade. Saturday was a working day, but the indefatigable Managing Director in whose hands were the accounts and moneys was at home indisposed.

Nobody noticed anything unusual about Archdeacon's behaviour on the Monday following when Captain Davies, being still unwell, gave the key of his desk to a relative and asked him to fetch certain papers. It was natural enough for the accountant to help this gentleman, who was not a member of the firm, in identifying the papers required. It is probable that Archdeacon saw two duplicate keys of this desk and, in the desk, a key of an inner compartment of the safe which contained securities. He, Archdeacon, already possessed the key to the outer door of the safe, and it is possible that he took the opportunity to pocket one of the desk keys. In any case he went out and once more drank himself into the mood of feeling that he had a right to better his financial condition.

At 11.30 next morning, Wednesday 14 January, Captain

Davies returned to the office. He found both his desk and the safe open. British securities worth about £12,000 were missing, also twenty-four $1,000 American bearer bonds together worth £5,000, a bag which contained £40 in gold and silver and two crossed cheques, one for £400 and the other for £12 10s 0d. The larger cheque was found crumpled up in the wastepaper basket, but apart from this all the easily realizable assets of the Company had disappeared. So had Archdeacon.

By 2.30 that day Captain Davies had mustered a quorum of Directors who passed the resolution already quoted. The police were called in and a reward of £300 offered for the apprehension of the thief. The police immediately sent Detective Sergeant Funnel to Liverpool, convinced that Archdeacon would take the first boat for the United States to cash the bearer bonds. Detective Sergeant Funnel drew a blank in Liverpool, nor was Archdeacon to be found in London. His lodgings in Alma Terrace, Kennington, were visited. The owner, Robert Gibble, a whitesmith, said that Archdeacon had come in between midnight and one o'clock the night before, very much intoxicated. He had thrown on to the mantelpiece some papers which he said were worth many thousands of pounds. Mr Gibble had thought this strange, but having known Archdeacon as a trustworthy lodger for six months he had not liked to say anything; and a few hours later Archdeacon had left the house taking the papers with him. This information was brought back to the Directors in the office in Cornhill.

The next thing to happen was that a cabby came to the office with a bundle of British securities worth £12,000. He said they had been left in his cab by a drunken fare.

But of the U.S. bonds, the cash, and the criminal there was no word for three more days. Then Archdeacon's brother-in-law visited Cornhill. He said that Archdeacon was waiting in his house to be arrested. The Managing Director, the Company's solicitor, and Detective Sergeant Funnel took a cab, picked him up, and handed him in at the Head Office of the City Police in Old Jewry. Archdeacon had on him £9 in cash, some postage stamps, and the receipt for a registered envelope addressed c/o Mr Cookson, the Post Office, Havant, Hants – which proved to contain the American Bonds. Archdeacon

c

said that he was very sorry. He had been drunk for a week but as soon as he became sober he had given himself up.

Laurence Archdeacon must have had endearing qualities. When he was brought before the Lord Mayor at the Mansion House His Worship remanded him for a week, remarking that he did not want the prisoner, who had evidently been drinking, to be at a disadvantage. And when the case was finally heard at the Old Bailey the Exchange Telegraph Company asked through their counsel that justice should be tempered by mercy. In his judgment the Recorder said that, when a person who was in a position of trust abused it and plundered his master, the duty of the Court unquestionably was to pass a sentence of penal servitude, but believing that in this case there were circumstances which would justify a milder punishment, he sentenced the prisoner to eighteen months' hard labour–which happened to be exactly the period Archdeacon had worked for the Company.

At the next Board Meeting it was resolved 'that in future the securities of the Company be deposited at the Bank, and that the Bankers be instructed only to give them up on receipt of a letter signed as in the case of a cheque, viz. by two Directors and countersigned by the Managing Director'.

The rest of the financial year passed quietly, and quite successfully. A ten per cent. dividend was paid on the A shares. As a result of an Extraordinary General Meeting in June the capital had been increased from £200,000 to £225,000 by the creation of 2,500 B shares at £10 each. The New York Stock Exchange and Paris Bourse prices were added to the service, and arrangements were in hand for including those of Berlin, Vienna, and Constantinople also. In the Annual Report it was stated:

> The improvements recently effected in the Company's instruments, by which the speed is more than doubled and a much greater security insured, admit of these services being given over these without prejudice to local matter.

As a result of Seyton's enquiries it appeared both possible and desirable to introduce the call system of the American District Telegraph Company as an aspect of the Company's business. This year, also, Deloitte, Dever, Griffiths and Co. became the Company's accountants. The name Dever has since been changed for Plender, but the firm has audited the Company's accounts ever since – now at a higher fee than the fifteen guineas then charged.

On 14 October 1874 the call system was introduced into the Stock Exchange and 'appeared to give satisfaction.' This was the arrangement by which subscribing firms could send through Bartholomew House messages to their representative on the floor of the Stock Exchange, and representatives could similarly get in touch with their own offices. This considerable addition to the stocks and shares service continued until the old Stock Exchange was closed on 6 February 1970.

The Company's service of stocks and shares prices from foreign capitals was by this time working well. One finds a letter from the Managing Director to the Editor of *The Daily Telegraph* on 1 July 1875 pointing out an error by that newspaper in its reports on Turkish securities. The Editor published the letter with a sour little comment which showed that he was offended by this interference.

There was a stir at the General Meeting in August of that year when a Mr Goslett proposed and a Mr Fearon seconded a motion 'that the meeting views with disappointment the heavy items charged in the account for the Directors' remuneration considering the same to be disproportioned to the result of the year's working, and the Directors are hereby requested to take such steps as they may think desirable for the reduction of such items in the future.' The motion was carried.

The point had been raised by Andrew Wark, one of the Directors, at the last Board Meeting. The total of Directors' fees amounted to £600 against a total for salaries and wages of £2,722 12s 1d. (One wonders who got the penny.) In the next year the Directors' fees fell to £142, less £50 returned by Charles Seyton. The biggest expenditure item was £800 for the New York and Paris telegraphic service. The Exchange system that year brought in over £7,000, but the call system which had been in operation for seven months, only £14 15s 0d.

The Company's activities were increasing slowly but steadily. A West End office, connected by an underground private Post

Office wire with Cornhill, was rented at 8 Piccadilly and 'the Company entered on possession' on Christmas Day 1876. But the usefulness of this branch office was limited by the extreme difficulty of obtaining wayleaves for overhead wires. The engineers would work gradually forward from roof to roof, only to be brought to a dead stop by some obstinate owner.

The Metropolitan Board of Works and Captain Shaw of the Fire Brigade became very interested in the Company's projected call system by fire boxes in the streets. (A type of fire alarm nothing to do with the Stock Exchange call system.) A model of this electric warning service was made for their inspection. Captain Shaw was favourably impressed; but orders had to come from the Metropolitan Board of Works, forerunners of the London County Council, and they were notoriously mean over equipment for the Fire Brigade. The Glasgow authorities, however, demonstrated that the traditional tight-fistedness of the Scot belongs more to funny stories than to fact. They placed an order for one hundred and twenty-five instruments at £6 each, plus 10s a year for maintenance.

A drive was made to induce newspapers to take the Exchange Telegraph's stock market service, but only *The Times,* the *Standard,* and the Central News Agency responded. With the Stock Exchange itself there remained certain difficulties. On Good Friday a new call indicator was installed. But in May one reads that it 'was removed and painted another color (*sic*) to suit the views of the Managers'. Patience was paying, however. The Fifth Annual General Meeting (1876) was told by Sir James Anderson, who had become Chairman:

> The Directors are now able to state that the position of the Company on the Stock Exchange, which up to the end of last year was one of sufferance, has now become recognized by the Managers, subject to certain conditions and the payment of an annual royalty.

Tiresome little troubles still remained, however. The Managers requested that the Company's man should leave the House by 4 p.m. A long, polite argument started on this subject.

For the rest, Frederick Higgins was the member of the Staff

most often mentioned. He made a habit of asking for a rise – which was always referred to the Directors, who generally held the question over. On 20 March 1877 he read a paper to the Society of Telegraph Engineers on the Company's type printing system. It was 'very satisfactorily received'. In April the Directors were concerned because the improved single wire instrument on which Higgins had long been working had been pirated by the constructing engineers and sold to the Post Office. And in June Higgins was on the mat for 'showing so much indiscretion in the tone of his letter in reply to a desire on the part of the Directors that the Call instruments should be kept in better order'.

The Company's main efforts at this time were concentrated upon having their fire alarm system more generally accepted and in inducing clubs and newspapers to take Parliamentary intelligence in addition to Stock Exchange prices. But the 1877 Report was cautious and a little sad. The business depression which had started the year before was holding back advance, the West End office had not proved productive, and for economy the call system instruments were removed from Bartholomew House to the Company's office in Cornhill, 'the work being equally well done'. The call system had so far proved a disappointment. The Chairman stated:

> The Call system has been pushed most energetically, advertised, lectured upon and exhibited without any appreciable result, and as nearly the whole of the sum expended in excess of called-up capital has been upon this system, the Directors consider it expedient to with-hold any further expenditure thereon for the present.

£710 had in fact been spent that year.

The business depression continued throughout the following year (1878) to the detriment of the Company's progress. A number of instruments were removed from subscribers who had gone bankrupt or failed to pay the fee. All increases of pay to the staff were held over, the total of Directors' fees was only £163, and the Company borrowed £300 for its immediate needs. In fact, apart from developments in the fire alarm system, the only innovation of the year was the hiring of three messenger boys at 9s a week. They were 'guaranteed for £10 by the Guarantee Society at a premium of 6s a year' – presumably against running away with their uniforms.

That messenger boys were engaged at Directors' level underlines the smallness of the London staff. There were the linesmen of whom we know only the name of the foreman, Bishop. There must have been electricians to care for the batteries and someone permanently in the West End office; no doubt there was a clerk or two, and another accountant had taken the place of Laurence Archdeacon who was still languishing (or getting drunk) in prison; but apart from the Managing Director the only employees ever referred to by name were Higgins the engineer, and Herbert, the reporter, who was installed when Christie unaccountably refused to go any more to the Stock Exchange. Numbers 17 and 18 Cornhill are now part of a massive pseudo-classic building dominated by Lloyds Bank. They were much smaller then; yet one wonders how the staff filled them. Typewriters had just been invented in America (It was some time before they crossed the Atlantic). But female typists were still a long way below the horizon – as were all lady employees. A woman's place was in the home, certainly not the City. The picture we have, then, in the middle of the first decade is of a very small Exchange Telegraph Company, widely extended, with a turnover of only about £7,000 a year.

But there is no doubt that it was a lively young company. It was continually developing new apparatus, buying patents, circulating various branches of business and directly canvassing contracts, trying by every means to grow. Both Directors and Staff accepted salaries which were small even by the standards of those days. The logical explanation of the comparatively large office accommodation is that the Company was determined to expand.

There was another slight brush with the Stock Exchange in December 1878. The Company proposed to issue a daily publication entitled *The Exchange Telegraph Recorder* which would list the prices of stocks and shares, much as the newspapers do now. To this the Committee for General Purposes of the Stock Exchange violently objected. Andrew Wark, a Director of the Company and also a member of the House, was thrice called before the Committee which 'urged the propriety of abandoning the project'. The Secretary of the Managers wrote to Captain Davies in the same terms. The Company's solicitors, being appealed to, advised that the proposal would be *ultra vires,* and it was therefore dropped.

Another hint – though not proof – of difficulty was that the Company's second reporter, Herbert, was at this period on the mat before the Chairman for the thoroughly unsatisfactory nature of his reports. But after giving his explanation (of which there is no record) he was retained.

Apart from the Chairman and the Managing Director, the Board at this time consisted only of E. W. Andrews, C. S. Seyton and Andrew Wark. At the beginning of 1879 Sir James Anderson felt it expedient to give up his honorarium as Chairman, which amounted to £100; and the Annual Report stated six months later that the business depression 'continued to exercise an adverse influence over the receipts of the Company'. But there were signs of improvement, largely due to the persistence with which the fire alarm system was pressed. Tenders for constructing the instruments were called for, and the lowest – £4 18s 0d each by Messrs Varley – was accepted. Captain Shaw of the Fire Brigade was becoming increasingly interested, and accepted an advertisement of the Company's system for inclusion in his book, *Fire Protection.* This is worth quoting almost in full for the description it gives of the service:

THE EXCHANGE TELEGRAPH COMPANY, LIMITED
   This Company is now establishing Branch Offices or Call stations throughout the Metropolis and Suburbs for the due administration of its
TELEGRAPHIC CALL SYSTEM,
By the aid of which, subscribers may be enabled any hour of the
DAY OR NIGHT
To call a
MESSENGER, CAB, or POLICEMAN,
and give the
ALARM OF FIRE,
While many other 'Calls', indicating the wants of a private

house, chambers, office, or place of business of any kind, may be arranged for, all such being made in the same uniform manner by the pressure of a button on a small automatic instrument, placed as most convenient, and telegraphically connected with the nearest 'Call Station' of the Company, which in no case will be distant from the subscriber more than a quarter of a mile, or three minutes' time. These instruments occupy but a few inches of space, are not liable to get out of order, require no local batteries or winding up, and no knowledge whatever of telegraphy to work them.

## CALL STATIONS

Will be established wherever a demand for them may arise; they will be provided with a permanent staff of Operators and Messengers, whose duty it will be to receive and attend to 'Calls'. A policeman will be found there, and an expert with a hand-pump or extincteur ready to act on the first alarm of fire, and each Station will be in telegraphic communication with the nearest

## POLICE AND FIRE BRIGADE STATIONS,

Thus enhancing the public value of the system by the increased security which will be rendered to life and property.

The system has been in operation in the United States for some years past, having been initiated and established by the American District Telegraph Company to meet a great public want, and where its practical advantages have become so fully recognised and appreciated that it is being rapidly adopted throughout the States; in New York, where it originated, and some thousands of instruments are at work, it is considered indispensable as

## THE HOME TELEGRAPH
## A PROTECTION TO LIFE AND PROPERTY – A CONVENIENCE IN DOMESTIC LIFE, AND AN ADJUNCT TO BUSINESS.

## DAY SERVICES,

Chiefly for the purpose of utilising the Messenger Service, will be established in business localities.

TERMS FOR A SINGLE INSTRUMENT:
Day and Night Service   ..     ..   5 Guineas per Annum.
Day Service     ..    ..    ..   4 Guineas per Annum.
      Extra Instruments at a Reduction.

Instruments placed and connected with the nearest Call Station of the Company, and kept in working order free of all charge to subscribers.

Messengers when employed, Sixpence to Eightpence per hour.

Automatic, Fire, and Burglar Alarms fitted in conjunction with the Call System.


Captain Eyre Massey Shaw, first Chief Officer of the London Fire Brigade, deserves further mention both as an ally of the Company during its first and difficult decade and also in his own right as a colourful personality. He was an Irishman of the family which later produced Bernard Shaw. He was intended by his father for the Church but escaped to America rather than take Holy Orders. This was, it is said, the only occasion when he ran away. He returned to join the Cork Rifles, and in 1859 – then a Captain aged twenty-nine – left the army to become Chief Constable and Superintendent of Fire Services in Belfast. With courage and tact he combated both fires and Protestant-Catholic outbreaks. When James Braidwood, Superintendent of the London Fire Brigade Establishment, was killed by a collapsing wall in London, the Establishment became the Fire Brigade with Captain Shaw as its commander.

He held this post for thirty years. He was a tall, aristocratic-looking man, a figure in society, a close friend of the Prince of Wales. He must have been widely known. The lovelorn Fairy Queen in Gilbert and Sullivan's *Iolanthe* sang of him thus:

Oh, Captain Shaw
Oh, Captain Shaw
Type of true love kept under,
Could thy Brigade
With cold cascade
Quench my great love, I wonder?

But in spite of the fine and varied qualities of its Chief Officer the Fire Brigade was not as efficient as it might have been, largely because of the parsimony of the Metropolitan Board of Works. Shaw, starved of the equipment he needed, saw in the Exchange Telegraph Company's call system a comparatively cheap means of getting early warning.

An agreement at this time was made with MacMahon's Telegraphic New Company for the use of the Exchange Telegraph's instruments to deliver 'general intelligence'. A licence was obtained from the Postmaster General, who had to be approached with regard to almost every new project. It was the first venture into the field of general news. This was hopeful. But in January 1880 the Company had to borrow £1,500.

The tide had turned, however. The first sign of this for the staff was that most of them received an increase of salary – or a rise as it was already called. Higgins was asked what he would accept in exchange for an undertaking to work exclusively for the Company and also to grant them the patent of any mechanical improvements or inventions he might make. He replied that he would be content with a salary of £450 a year. When an agreement to this effect was signed he was also given for his patents £500 in cash and 200 B shares.

Another sign of progress was that before the end of 1881 a telephone was installed in the office. The minutes state that the Manager of the United Telephone Company was invited to use this instrument when he wished, to test its 'utility.'

A few Board Meetings later the Directors appear to have lost their heads, for they go on record as purchasing one hundred telephones. But this action is explained by a circular dated January 1882. The Directors 'after careful consideration of the subject decided to give subscribing members telephones to enable them to communicate with the Company's Office, and with each other, and later on with the Stock Exchange, if permission can be obtained from the Managers.'

The Managers had already been sounded without success. The Stock Exchange remained prickly. The reporter, Herbert, had been criticized for having both his clerks in the House at once. It is interesting that at this time negotiations were started to rent two rooms in Bartholomew House, the building which had formerly been occupied by the stockbroking firm of Borthwick and Wark – one a former Director and the other still at that time on the Board. It will be remembered that their office had for a short while been used for the Company's business, but this new move was the beginning of 'B.H.' as a telephone call exchange run by the Company.

The year 1881 was busy and expansive. The Company's instruments were shown at exhibitions in Paris and at the Crystal Palace where a gold medal was won. At the General Meeting 'substantial improvements in business' were announced, and a ten per cent dividend – less income tax at 5d in the pound.

No mention was made of the ten years anniversary although there was a Board Meeting actually on the 28 March 1882. But there were two important announcements. The Glasgow Exchange Telegraph Company and the Liverpool Exchange Telegraph Company were to be launched. And in July the MacMahon Company, in conjunction with which the first general news service had been started, was taken over by the Exchange Telegraph.

The Company's turnover was £15,000. It was not yet rich. But it was thrusting out in all directions and had proved that it was among the first to try something new.

CHAPTER 3

# The Second Decade

IT WAS at the dawn of its second decade that the Company began to provide 'general news', which included some sport, not only to London subscribers but also to those in the provinces through the Liverpool, Manchester and Glasgow offices. A folder entitled *Tariff for 1883* sets out the services in detail.

The Exchange Telegraph Company, Limited, has for ten years supplied direct from the London Stock Exchange, by means of its recording instruments, the quotations from which the News of business done has subsequently reached the Exchanges, Newspapers, and Clubs in London and the Provinces. MacMahon's Telegraphic News Company, which was incorporated with *The Exchange Telegraph Company* in July last, has, in the same way, distributed General and Racing News to all the Chief Clubs and other Institutions in London. In extending its operations to the Provinces, the Company has, therefore, an organization adapted to the work it undertakes; and with regard to Commercial News, it has acquired a prestige for speed and accuracy unsurpassed by any other Agency.

*The Exchange Telegraph Company,* on its enlarged basis, had, as one of its first operations, to organize Agencies in Egypt for the purpose of transmitting, as early as possible, news of the important events then impending [the war with the forces of the Mahdi which later led to the murder of General Gordon at Khartoum]; and its reports of the principal incidents of the campaign were, with hardly an exception, the first that were received in London and the provinces. The same enterprise will distinguish the Company in all its arrangements abroad. Agents have been appointed in most of the chief cities of the world, and have been instructed to forward by telegraph news of every important and interesting event. At home the United Kingdom is already well covered by its Corresponding Agents.

The Special and other Telegrams of *The Exchange Telegraph Company* will, after the first of January 1883, be received only by the Exchanges, Clubs, and Newspapers subscribing directly to the Company for the services of News comprising such telegrams. No other Agency will be privileged to distribute the telegrams of the Company.

The Directors of *The Exchange Telegraph Company* wish it to be well understood that their arrangements are made chiefly with a view to the supply of News during the day and evening to Exchanges and Clubs and to Evening Newspapers, and to these they have every confidence in recommending the various classes of News and the Market Reports, etc., quoted in the accompanying tariff.

### GENERAL NEWS

|  | Monthly Charge |
|---|---|
| THE MORNING EXPRESS. CLASS I. – Giving a full epitome of the London Morning Papers, dispatched between 4 and 5 o'clock a.m. daily .. .. | £ 2 12 0 |
| CLASS II. – About three hundred words daily .. .. .. .. | 1 6 0 |
| CLASS III. – The most important items dispatched at 6 a.m. .. .. | 0 13 0 |
| COMMERCIAL MORNING NEWS from the *Times* .. .. .. | 0 13 0 |
| MIDDAY NEWS – A dispatch at noon, or in the course of the day, between 9 a.m. and 3 p.m., summarising the News which has transpired up to the time of telegraphing .. .. .. | 0 13 0 |
| GENERAL NEWS CLASS I. – Giving all the important News which transpires from 9 a.m. to 6 p.m., including Foreign Telegrams .. .. .. | 3 0 0 |
| The same to 3 p.m. .. .. .. | 1 10 0 |

CLASS II. – Giving the more important portion of Class I., two or three
Telegrams daily, between 9 a.m. and
6 p.m...    ..    ..    ..    ..        1   6   0

The results of the Derby, Oaks, Two Thousand
Guineas, Ascot Cup, Chester Cup, St. Leger,
Cesarewitch, Cambridgeshire, The Great Boat
Races, and the Winners of the Waterloo Cup,
Purse, and Plate are included in General News,
Class I and II.

EVENING NEWS, from 6 to 10 p.m. ..    ..       0 15   0
SUNDAY NEWS, from 9 a.m. to 9 p.m.    ..       0   8   8
PARLIAMENTARY NEWS. – A summary of
the proceedings from the opening of the House
to 10 p.m. Suitable for Clubs ..    ..    ..       2   5   0

PARLIAMENTARY NEWS from the opening
of the House to 6 p.m. Adapted to the requirements of an Evening Newspaper in the Provinces    ..    ..    ..    ..    ..    ..       2   5   0

These charges are only made during the
months in which Parliament sits. No deduction
is made during the Easter and Whitsuntide
recesses.

The folder goes on to give the details and cost of Stock
Exchange news, and daily market reports – butter, corn, cotton,
iron, sugar, wool. Wherever the Company had no private wire
the information was transmitted by telegram at press rates.
Considering the modest charges quoted there cannot have been
much margin of profit. A great deal of money was being spent –
to the vocal disquiet of some of the shareholders.

But it was necessary to enlarge and modernize to keep ahead.
Rivals were beginning to appear. In the circular to shareholders announcing the take-over of the MacMahon Company
for £8,500 in cash and eight hundred fully paid B shares this
information was also given:

The Directors take this opportunity of referring to a
Prospectus recently issued of a Company, calling itself "The

Electric News Company Limited" organized to purchase the Patent Rights and Business of the Automatic Telegraph Company Limited, a Company which for about fifteen months has been making vain attempts, by reducing the subscription to an unprofitable point, etc. to oppose the MacMahon and Exchange Companies. In their published Prospectus they claim to possess a superior instrument and valuable patents, both of which claims we do not admit; but, on the contrary, we maintain that we have patents, both old and new, for much superior instruments to any in this country.

This new company, however false its claims, continued to bleed the Exchange Telegraph of its clients – so much so that in June 1883 the following Notice to Subscribers was sent out.

Our attention has been drawn to a card and circular issued by the *Electric News Telegraph Company,* offering to *new* subscribers an instrumental service of Stock and Share quotations for £20 per annum, and "three months free to parties giving up the other Company's instruments."

We can assure you that it is quite impossible to permanently supply an efficient service at this rate.

This action on the part of the *Electric News Company* is an attempt to take away our Subscribers, and is in consequence of our refusal to accept an arrangement proposed by that Company and involving an increase of rates.

We trust this explanation will be satisfactory to you, and that in the interests of a permanently complete and efficient service, which we shall always endeavour to maintain, you will continue to support this Company as hitherto.

By order of the Board

W. KING
*Assistant Secretary*

There are two points to notice here – the fighting tone and the name of the Assistant Secretary who signed the notice. Presumably he also drafted it, for the only other person who might have done so would have been the Managing Director who still also acted as Secretary; and it cannot have been he, for in all the many compositions of W. H. Davies which have been preserved there is not one split infinitive. But here is someone too angry to bother about syntax or at least liable to this grammatical foible.*

W. King was first mentioned in the account books early in 1882 as receiving a wage of £4 9s 10d a month. Shortly afterwards he had been promoted to Assistant Secretary at £12 10s 0d. It is appropriate that this Notice to Subscribers should, so to speak, be his first public appearance.

The Company had already had a brush with the Central News which began to offer 'general, parliamentary, and sporting intelligence and other matters connected therewith' in addition to Stock Exchange prices. With techniques of communication continually improving such rivalry was likely to increase. But in one instance only were the Directors ready to accept a compromise solution. E. B. Bright, elder brother of Sir Charles Bright, the electric engineer responsible for the laying of the first (unsuccessful) Atlantic cable, and for other major projects, was operating a fire alarm system which was at least as good as that of the Exchange Telegraph. Finding that too much time and money would be spent in active competition with this, it was decided to form a separate concern – the Electric Fire Alarm and Signals Company – in conjunction with Bright, and to sell their patents to this company. In the event, however, it was found impossible to float the company, the application for shares being insufficient; so the two fire alarm systems continued to compete.

All in all it was a difficult year, and the Annual General Meeting must have been a melancholy occasion. The Directors announced no dividend, a net loss of £1,298, and the resignation of Sir James Anderson.

On the going of the father of the Company, no new Chairman was for some time elected, Henry Burt, J. H. Hutchinson and C. S. Seyton taking the Chair in turn. No member of the Board stood out sufficiently as a personality to step into Sir James's shoes. Also, although the old sailor was still very much alive

---

*Soon afterwards Wilfred King wrote that the Directors were 'shortly to thoroughly consider the question of consolidation of capital.' And also that they will not fail 'to again bring forward' the same matter.

there was in the Cornhill office a feeling like mourning, such as forbids the appointment of a new spouse for at least a year.

This last statement is only deduced. In contrast there was a sense of urgency, plenty of the fighting spirit. There was need of it. During the following months an action with the Central News was fought and won. But the Central News appealed for a re-trial, and although there seemed little prospect of this being granted, a compromise was finally agreed out of court, each side paying its own costs. Those of the Company were £1,521. This, coupled with continued fishing for the Company's subscribers by the Electric News Agency, using the bait of lower rates, resulted in the Annual General Report for 1884 showing a debit of £6,863, the worst to date – and still no dividend, of course.

There followed an economy drive, with which the new Secretary, Wilfred King, was mainly concerned. But that alone could not cure the evil. The Board in effect decided to act on the motto, 'If you can't beat 'em, eat 'em', and set about engulfing the Electric News.

The negotiations were long and difficult, carried on with Henry Burt in the front line and the Board backing up or instructing. An early suggestion was to rent the offending company for £1,600 a year. But again and again they inserted 'objectionable clauses' until it was decided that agreement was impossible, and an all-out assault was made with the object of buying the business outright. It was a bold plan while the Exchange Telegraph was deeply in the red. But the Electric News was in still worse state, having irremediably weakened itself by the continued policy of undercutting, and at last they agreed to the take-over in exchange for £17,000 in £10 shares credited with £8 paid. The Exchange Telegraph increased its capital by 2,125 A shares, valued at £17,000, and one of the Directors of the deceased company, Sir Michael Kennedy, joined the Board. The Agreement was signed and sealed on 14 August 1884. It was Higgins who actually handed over the purchase price, and 'took over charge of the Electric News working the following morning'.

Henry Burt received a vote of thanks for his patient work as negotiator and a cheque for £100. It may only have been coincidental but Wilfred King at this time had his salary raised

D

from £12 10s 0d to £16 13s 4d a month. Active as ever, he sent
out a polite but firmly worded letter to the subscribers of the
Electric News, asking them to decide within seven days whether
they preferred to continue with the Exchange Telegraph or to
have their instruments, which were now the property of
the Company, removed. He was no doubt also concerned
– although there is no signature – with a cock-a-hoop prospectus which began with the long sentence, 'The system of distributing news adopted by the Exchange Telegraph Company
in London and the three Provincial Towns in which it has
established centres has secured for the Company a practical
monopoly for the supplies to the Exchanges and Clubs in those
places, no other agency being able to give, for a reasonable sum,
anything like the quantity of news furnished by the Exchange
Company, or to circulate the information with the same promptitude'.

Attached was a comparison of the annual charges made by
the Company and by the Central News: 'Morning News (three
hundred words): Exchange, £12 12s 0d: Central News, £15 12s
0d: Difference, £3 0s 0d . . . etc.' Sleeves were being rolled up
to deal with the next competitor.

Two other incidents of this year may be recorded, one satisfactory, the other not. Arrangements were made for private
wires to Lord's and to the Oval, although in the latter case it was
not found possible to carry the wire within the grounds, and a
room overlooking the cricket field had to be taken.

It was discovered that a good deal of money had disappeared
from the Manchester office, and was unaccounted for. The
Manager, Mr Butler, was given until the following Saturday to
make good the deficit or be charged with theft. The Company
suffered from this clemency. On Saturday the money was still
short and the Manager had disappeared, reportedly overseas.

This was only the second act of dishonesty in the Company's
history, but it evidently hurt. The Exchange Telegraph was a
family firm, not in the sense of nepotism – far from it – but it
was small enough for everyone to know everyone; they were all
hand-picked from the Managing Director to the messenger
boys, and once appointed they tended to remain throughout
their working lives. Captain W. H. Davies was no doubt the

nominee of Sir James Anderson, for Davies had been first officer of the *Great Eastern* on her maiden voyage to New York. He served as Acting Secretary until he proved his ability to steer a young company as surely as a ship and to look after the personnel. He evidently had a tremendous capacity for work. Question after difficult question was regularly referred to him for report or action at each Board Meeting, and he always had a full report to give a fortnight later. One journalist who interviewed him described him as 'jovial Captain', and his portrait suggests a cheerful man of the sea. The eyes are intelligent. He certainly kept the staff working, but he must have been popular, for the employees stayed. Only Christie, the first financial reporter – who was evidently allergic to the Stock Exchange or he to it – left of his own accord, and Laurence Archdeacon of necessity. The rest stayed – scarcely for the money. It is difficult to gauge the value of contemporary wages. But between 10s and £1 a week cannot have bought much more than bread and butter for a clerk with a family. Messengers received less than 10s a week. On the other hand the minutes tell of £14 being spent on one occasion for their uniforms and £21 for their boots. They, unlike Napoleon's army which marched on its stomach, ran errands on their boots. They were a smart, even élite little corps. No doubt they misbehaved as healthy boys do. But firms on whom they called often tried to seduce them away from the Exchange Telegraph. And – this is the important point – many of them, and junior clerks too, remained with the Company to rise eventually to senior posts. There are still senior members of the Company who started as messenger boys or office boys at less than 10s a week.

Their salaries were slow to rise. During the depression years of the early 1880s there were no rises at all. The whole staff accepted this. When at last this squeeze was relaxed eighteen employees received increases totalling £1 7s 0d a week – less than 1s 6d each. There was little inducement to stay apart from loyalty and belief in the Company's future.

The events of the mid-1880s may be summarized. In 1884 the debit balance was £6,836, in 1885 £1,465. At the General Meeting of the latter year the Directors reported a marked increase of business and 'the prospect of a resumption of Dividends at no distant date'. Revenue had attained a rate of nearly £37,000 a year, there being over four hundred subscribers to the financial service and about half that number to the general news. Every effort was being made to keep down expenditure.

What might have been a major crisis arose when the governing body of the Stock Exchange 'considered it undesirable to continue their connection with the company and gave the necessary notice to terminate the Agreement'. The Company at once removed its telegraphist and wires. But they did not transport them far, only to the predecessor of the present Bartholomew House, and the reporter remained. Consequently the shareholders could be told: 'The business of the Company has in no way been prejudiced thereby, on the contrary the Directors have been able to give subscribers on the Stock Exchange further facilities for doing business that are much appreciated'. This was a telephone, just outside the House.

New York stock exchange prices had, of course, been provided for some time. In 1885 the Board decided 'to give the Dow Jones system a trial for one month'. This had nothing to do with the Dow Jones Average which we hear about today. That was not in operation until 1896. It meant that the Dow Jones firm were to act as the Company's correspondent in New York.

Although business was better, there was still an accent on economy. The stated aim was to reduce expenditure to £25,000 a year. This was not achieved, but every possible saving was made. Directors' fees were fixed at £50 basic for a year, plus up to £100 dependent on attendance. There was a lot of Boardroom discussion on the desirability or otherwise of reducing foreign news. It was much the most expensive item of the 'general intelligence'. But at this time – when Khartoum was lost and recaptured – it was felt that no economy could be made here. When peace returned, the contract for general news specified that the fee would be automatically increased by twenty per cent. in the event of war.

An interesting point about these general news contracts, which included racing, was that the information was for the use of the recipient only and was not to be passed on to a third party. Since the recipient was as often as not a club this sounds a little absurd. But, although the Company was always well on the right side of the law it knew that its subscribers often steered very near to this inhibiting wind, and sometimes against it. In Manchester in 1885 twenty-four clubs were temporarily closed following a police raid. There was trouble at Kempton Park, and a messenger was dismissed for giving information to a loafer. In February 1886 the instrument at the Somerset Club was seized by the police. The Board consulted their solicitor 'who deprecated legal proceedings, the law being against us, and recommended conciliatory measures'.

The solicitor was often called in to Board Meetings at this period. There was a variety of troubles of which only a few examples need be given. Twice within a twelve-month it was noted in the minutes that a working arrangement had been arrived at with the Press Association – which shows there was already trouble, though only of a minor nature. Then there was the curious case of Myer Myers.

This gentleman called several times at the Company's Haymarket office in September 1886. He saw the manager, John Gregory, and asked him if this was the office from which the names of the horses were sent out. He was told that it was. He asked if he could have an instrument fixed in his Bishopsgate office. He was told that he could, for an annual rental of £50. Myers then told Gregory that he was the publishing agent of the General Steam Navigation Company, that he had written two books about it, and that he could obtain passes for any destination.

A few days later he called again, and gave Gregory two passes for Edinburgh. He remarked, 'You and I could make ten guineas a week easily'.

'How?' Gregory asked.

'By holding back racing news until I have received the name of the winner and betted on it,' was the reply.

Perhaps Gregory did not react as forcibly as he should have done to this barefaced suggestion. He merely said that it was out of the question, and when Myers had gone informed the head office.

In November Myers called again, and again spoke to Gregory about his scheme, this time in the presence of a clerk named Fullex. A policeman was sent for, and Myers was arrested. Giving evidence, P.C. Balls stated that the prisoner had said to him, 'Do not lock me up. I am a ruined man'.

However, his defence was effective. His counsel pointed out that he was charged under an Act which covered the damaging and injury of telegraphic apparatus. Of this Myers was not guilty. He had merely suggested delaying a message, which was not covered by the Act.

Myers was remanded for a week on bail of £200 to allow the prosecution time to make up their minds what course to follow. They decided to proceed with the charge but to add another – inciting Gregory and Fullex to conspire with him.

This interesting case was never heard, for Myers killed himself by taking an overdose of laudanum.

The case of Henry Wilks is less dramatic, but it is worth giving *The Times* account of the trial almost in full for the incidental information it provides on how racing news was transmitted.

> Henry Wilks, twenty-one, a telegraphist, pleaded Guilty to an indictment charging him with altering and delaying the contents of certain telegrams . . .
>
> Mr. Paul Taylor was counsel for the defence.
>
> This was a prosecution instituted under sections of the Post Office Protection Act which make it an offence to alter, delay, or disclose the contents of a telegram. The prisoner was a telegraphist employed at the Central Telegraph Office in London. It seemed that it would not have been the prisoner's duty to have been at the place he was on the day in question unless he had arranged to take the place of another operator who was receiving messages. It was stated that the prisoner had made several previous attempts to be put in a position to receive the telegrams. At Sandown Park and other racecourses before a race takes place the numbers of the horses running are hoisted on a board in the order in which they appear on the race card. The number of starters is immediately telegraphed to the Central Office in London, and the message is forwarded to the Exchange Telegraph Company by whom it is instantly

transmitted to the different clubs in London on the tape, and betting then takes place. Immediately the race is over, the result and the names of the first, second, and third horses are telegraphed from the course to the Central Office, and are transmitted to the clubs in the same manner. On the day in question the starters for the third race at Sandown Park were telegraphed from the racecourse in the order in which they appeared on the card. On arriving at the Central Office the message was handed to the prisoner for him to send it to the Exchange Telegraph Company. The prisoner held the telegram for some minutes until he had ascertained what the winner of the race was, and he then telegraphed the starters to the Exchange Telegraph Company, putting the name of the winner in a particular place. In the race the winner was eighth on the list of starters, but the prisoner placed it third, so that the persons who were acting in concert with him knew by prearrangement that that was the winner. A few minutes afterwards the prisoner telegraphed the result of the race. Inquiries were eventually made, and on being spoken to on the subject the prisoner said he had been induced to do it by other persons who had made him drunk, and that he had not benefited by it. It was stated, however, that there was no foundation for the statement that he was drunk at the time. The system of check was so perfect that it was impossible for a matter of this kind to escape detection.

Mr J. P. Grain, who represented the Exchange Telegraph Company, said they took every precaution to prevent fraud being committed. They were simply purveyors of racing news to the public in the same manner as they supplied every other kind of news. They offered every possible facility to the Post Office, and they themselves prosecuted another man, who was sent for trial, but who committed suicide before his trial came on.

Mr Taylor addressed the Recorder on behalf of the prisoner, who, he contended, had been the dupe of other persons. The prisoner unfortunately became connected with some low-class clubs, and for a certain time resisted the solicitations of persons who endeavoured to induce him to become a party to the fraud. The prisoner had not benefited by the matter.

The Recorder said the prosecution might have been instituted summarily, when the prisoner would have been liable to a fine, but he thought the authorities were quite justified in taking a more serious view of the matter and proceeding by indictment. Having commented on the importance there was in preserving regularity in the transmission of telegraphic messages, the Recorder said he would take into consideration the prisoner's youth and the fact that no one could attempt to commit such an offence without the absolute certainty of being detected. The prisoner had been in prison awaiting trial upon the charge for six weeks, and the Recorder now sentenced him to one month's imprisonment.

A telegram case of a different sort also deserves mention. On 28 December 1886, the following news telegram was delivered at the Exchange Telegraph office in Liverpool:

> Stalls at Drury Lane Pantomime were empty at ten o'clock. Performance finished at one. Procession and scenes good otherwise great failure. Sala.

Sala was the famous *Daily Telegraph* reporter and leader-writer, the best-known name in Fleet Street at the time. The Liverpool office might have been surprised that he should send this telegram to them, but they accepted it and sent it out. When Augustine Harris of Drury Lane read it in his newspaper (not the *Daily Telegraph*!) he was greatly incensed and threatened Sala with a libel action. Sala denied authorship.

On 1 January Augustine Harris's solicitors wrote to the company. At the first Board Meeting of the year this letter was read, also the Company's letter to the General Post Office and their reply. (This correspondence is lost but one can guess at its content.) 'The matter having been duly discussed and the solicitor consulted thereon,' state the minutes, 'it was decided to await further proceedings on the part of Mr. Harris.'

They did not have to wait for long. The dignified and possibly post-prandial atmosphere of the 2.30 p.m. meeting was soon disturbed. Wilfred King, who by the handwriting wrote the minutes, threw away in a frigid sentence as dramatic a moment as was ever portrayed at Drury Lane: 'Before the Board rose, a writ was served on the Company in re the above and the Managing Director was instructed to place the matter in the hands of the Company's solicitors.'

The case ended in anti-climax five months later when Augustine Harris indicated that he was prepared to withdraw the charge in exchange for twenty guineas to cover his costs. He may have felt there was no profit in pursuing an incident which might have been a practical joke and in any case had been forgotten by the public. There was never any explanation. Signatures are always at risk to the impostor who uses telegrams.

As a final example of the need of recourse to the law, someone started cutting the wires which made a spider's web above the roofs of London. A reward was offered for information leading to the arrest of the culprit, and a man named Albert Saunders was arrested, convicted, and sentenced to three months' hard labour. His motive was not divulged.

But it was not rival organizations or human conspirators and saboteurs that caused most damage to the Company in 1886 and 1887. It was the English climate. During both these winters there were heavy snowstorms and high winds which not only broke many wires but damaged roofs and the chimney-stacks to which the poles were fixed. Between Boxing Day 1886 and 10 January following, the weather was appalling. The report written by the chief engineer, Frederick Higgins, deserves to be recorded:

On the evening and night of the 26 December 1886 a most destructive snowstorm occurred accompanied in this district by a slight gale and approximating in conditions to that described in par. 12 of my report of 20 January 1886.

The snow was of unusual density melting in the condition in which it adhered to our wires into about one-eighth of its bulk, the usual density being from one-eighteenth and over. The weight of the cylinder of snow thus supported by our wires would be about fifty pounds in one hundred yards for each wire, this added to the usual strain further augmented by wind and contraction rendered it almost impossible for the supports to bear the load, fortunately they in a large number of cases gave way without seriously injuring the masonry or buildings.

On the morning of the 27th the whole of our system was interrupted including the wires rented from the Post Office for our own system and Reuters.

The destruction was not so wholesale as this complete interruption would lead one to suppose because our circuits being mostly of considerable length were each interrupted at some point or other.

The work of restoring communication was commenced under circumstances of great difficulty and danger arising from further falls of snow, of telephone wires from thaw, fog, gale and rain in succession.

The whole of our City system was temporarily repaired with the exception of thirty-six instruments by Saturday morning 1st January but on Monday in consequence of a heavy fall of sleet the number of instruments interrupted was increased to two hundred and thirty-six.

Again on Tuesday the 4th although four hundred and thirty-four Financial instruments were working out of a total of four hundred and sixty-six only three hundred and eighty-one were workable next day.

On the 10th January all the Financial including our lines to our West End Offices were again at work with the exception of three subscribers in Holborn and twenty beyond the West End office.

Of the General News system there are at present twenty-one instruments at work.

The damage caused by the wires to old and delapidated chimneys has been very extensive.

I am unable to make a full report upon the nature of the damage already repaired and the extent of that remaining as it extends over some hundreds of miles of housetops and the accounts of the work already done will require some leisure time to analyse. The damage to chimneys etc. already ascertained exceeds £500.

From what I have seen I should think the damage is from six to ten times as great as that caused by the storm of January 1886.

On strong buildings the wires have given way and on weak buildings the wires are frequently unbroken and the chimneys down, thus shewing that any greater strength of construction would only have added to the mischief.

A number of our wires have evidently been carried away by the fall of telephone wires which are much too numerous overhead and not only render wayleaves difficult to obtain but compel the adoption of long and dangerous spans of wire.

Our cables appear to be unimpaired, nothing was disturbed on the Royal Exchange and our derrick on Bartholomew House was uninjured. One six inch oak post and five-eighths inch iron stay rod broke on this house under the strain of five cables but was secured. The work of restoration and repair will be equal to about half to two thirds of the construction of new lines.

I have at present sixty men employed in thirteen gangs which number cannot well be increased unless we obtain some skilled wiremen to take charge of gangs in which case the work will proceed more rapidly.

In spite of the energy with which the work was pressed forward, repairs were still being carried out in April. One reason given for this was that the Telephone Company, which was slower off the mark, pulled its damaged wires over the Exchange Telegraph's repaired wires and damaged them afresh. All in all it was a major operation which cost the Company £1,255 and the lives of two linesmen.

In addition to this the cost of wayleaves went up—the *Christian World* demanded £10 for a mast on its roof—and often they were refused altogether. Higgins wrote of one of his men who had 'worn his boots out' in attempts to get a wayleave.

Clearly the better policy was to go underground. But that was much more easily resolved than done, for there was so much already under the narrow city streets; and permission to dig up the pavements was hard to obtain. As a result, a remarkable minute was recorded: 'Resolved to memorialize the Commissioners of Sewers'. But although this solemn ceremonial was performed no useful concession was granted. So it was a case of getting under the roadway somehow. Here is what the picturesque Mr Higgins reported of Whitefriars Street:

> The street turned out when opened up to be full of pipes in all directions which at one place threatened to terminate the laying and render the part already laid useless. By packing very closely and encroaching somewhat upon a coal hole the pipe was continued and I think without bends which will obstruct the drawing of the wires.

By this time there were more than five hundred financial service subscribers—a lot of lines to look after. Business was going better. The improvement had begun—noticeably for the shareholders—at the beginning of 1886. The minutes of the first Board Meeting of that year noted with satisfaction that the dividend cheques signed on 31 December had all been posted on 1 January.

A number of innovations were made in that year and those immediately following. Political information was added to the general news. News sheets were put up in hotels, and instruments were supplied to the police for their better communication. Telephone boxes were introduced into the Stock Exchange—in the cloak-room. They were padded within with felt and lined with cloth. Higgins wrote:

> I think the boxes are sufficiently sound tight to prevent the communication being made in one box from being overheard in another but not to prevent a listener outside from hearing what was going on within if spoken above the ordinary conversational pitch.

But the achievement was to have got them there. Higgins only noted in passing that no partition had been erected to separate the telephone from the ordinary business of the cloak-room.

There had so far been no particular emphasis on the turf,

but a racing monopoly was established at Brighton. A guinea service was started by which a subscriber received a guinea's worth of telegrams. This was still active in the time of recently retired members of the staff. For the staff, subscriptions for a convalescent home were started and an annual dinner—the first of many—was held.

There were many innovations on the mechanical side. In place of batteries, accumulators charged by hydraulic power were installed. And Higgins invented a column printer. All in all it was his decade. He was always inventing something or reporting on a machine invented by another. In the latter occupation he showed a dry and caustic wit. For instance:

> The keyboard . . . is ingeniously contrived to prevent anything like fast manipulation even after the instrument itself can be got to work properly.

His work did not go unrewarded. In 1889 he received £1,000 in final payment for his patents, and his bonus was raised to £150 a year. But he was not always the darling of the Directors, for he did things in his own way and spoke his mind.

The editorial department was at this period quite often under criticism—for not being prompt enough, for inaccuracy, for incurring too large a telegram bill. The Editor, J. F. Andrews, sent out a circular to correspondents. (They were what journalists called stringers, men with other jobs who were paid—in this case—two shillings for each news telegram accepted.) Andrews guided them on the type of news required:

> Political news and information affecting public men, and also news of an important commercial character are wanted . . . but casualties of an ordinary type, police court cases, suicides, state of weather etc., *unless of a very remarkable character,* are never sent out by this Agency.

More than half the circular was concerned with saving words and therefore telegraphic costs. All telegrams should be addressed to Diocles, London.

> The ordinal numbers, if written 1st, 2nd, 3rd, 4th etc., are counted as two words; but first, second, third, fourth etc.

would of course count as one . . . Brevity should, as hitherto, be studied as much as possible, and where ten words will convey a fact no more should be sent.

Higgins was also under pressure to economize, which he evidently considered downright inhuman when it concerned machines. Asked for a return on the paper his department had used, he replied: 'The cost of paper was 17s 2d against 19s 1d per instrument per annum last year. The reduction would have been greater if a briefer form of expression had been used in the messages'.

Death touched the still youthful Company. C. S. Seyton, one of the most active of the Directors, died in 1886. Lord Borthwick, an ex-director, had died on the last day of the preceding year. Kenneth Anderson, the stockbroker son of the Founder, was invited to join the Board in place of Seyton. He accepted, and remained a Director until 1914.

In 1889 Henry Burt resigned from the Board from pressure of his other business interests. A paragraph from his good-bye letter to the Managing Director makes a suitable ending to this chapter:

It will always be a matter of satisfaction to me that my connection with the Company though commenced when the expenditure was in excess of income, should terminate after some years of anxiety and hard work at a time when the nett profit has nearly reached the sum of £13,000 per annum.

# CHAPTER 4

# The Turn of the Century

IN SEPTEMBER 1889 the Board had considered the purchase of 'a Type Writer' [*sic*]. They did not then reach a decision, but in the following month they resolved to buy. In March, 1890, a shorthand clerk was engaged at £1 a week. Four years later it was brought to the notice of the Managing Director that the column printer 'could not keep up with the Morse (financial news was sent in Morse) on account of there not being sufficient fractions'. Wyckoff, Seamens & Co. then suggested that Captain Davies should experiment with a Remington typewriter with a keyboard which included fractions. They leased the instrument with its own operator to the Company for 45s a week, cost of paper extra. The experiment was sufficiently successful for the Board to decide upon engaging their own skilled typist at 30s a week.

These clerks were of course men. The first woman to be mentioned—in 1895—was Miss Tomlinson, who is described as a typist. She did not last for long. She died in 1898, and the Board authorized payment of the doctor's bill.

These points may appear trivial, but they help one to form a picture of the Cornhill office—a company with an annual revenue of £48,000 (1889), with no typewriters and no women. Since as much as thirty years later dark suits were obligatory, one may assume that dress was strictly formal—frock coats and high collars. The clerks sat on high stools at counter-like desks (these remained well into the next century), and wrote, not with quill pens, admittedly, but with steel nibs which were issued with care as the valuable things they were. It is a Dickensian picture.

How much everybody must have written, from the Managing

Director down to the junior clerks earning about 10s a week! At every Board Meeting a number of letters were referred to or read—perhaps a dozen or more. These must have been a very small proportion of the total correspondence, plus copies. Penny-a-liner was the designation of a hack writer. These clerks earned much less than a penny a line. There is something to be said for typewriters and girls.

One is conscious of paradox. Thanks to the genius of their engineer, the staff of the Exchange Telegraph were familiar with such wonders of science as the tape machine and column printer. But until the end of 1889 they did not know the typewriter. This is comparable to the Eskimo who became accustomed to aeroplanes long before he saw a motor-car.

Higgins remained as full of ideas, energy and caustic humour as ever. With his passion for statistics he began giving in his annual reports not only the number of instruments in use but the number of words that they transmitted. By 1898 this total was 317 million. He calculated that if the messages had been sent out at Post Office telegraphic rates the cost would have been £158,000.

The Stock Exchange telephones, he claimed, were proving thoroughly satisfactory. Connections were made so quickly that brokers could often be seen leaving a box fifteen seconds after entering, having spoken with their office in this time. The number of quotations sent out on the tape ran into hundreds of thousands, and the system of naming stocks in which brokers were anxious to deal was increasingly popular. Nearly 500,000 calls passed through Bartholomew House. But the most outstanding success was the column printer, constructed in the Company's own workshop which had just been set up. In 1890 only four of these were in use. By the next year the number had risen to forty-six, and by 1892 to seventy-four, with twenty-six spares. These instruments could deliver 2,000 words an hour. Reuters started a news service with eight of the Company's column printers and seven of the column printer tape instruments, and the Press Association tried to make a similar agreement soon afterwards. Higgins was now working on a magneto instrument.

For some years there had been an annunciator in the smoking-room of the House of Commons. This told in large type who was speaking, when there was a division and so on. By 1895 the number of annunciators about the House had, at the request of Members, been increased to three. Higgins reported, 'They are not likely to be much increased as they are supposed to keep the Members from attending the Debates'.

Although attempts were continually made to lay cables underground the Commissioners of Sewers remained deaf to all appeals, so the roofs of London were still strung with a network of wires. Higgins constantly complained of other people's bad wires falling on his good ones. The authorities became alive to the situation to the extent that they passed by-laws which stipulated certain precautions, and they sent inspectors round to see that these were taken. Higgins did not think much of these people. He reported:

> The new inspectors of overhead wires give as much trouble as they possibly can in a variety of ways and not being practically acquainted with the subject involve us in a lot of correspondence. For instance a pole was said to be 'not efficiently stayed as required by the bye-laws'. This was literally correct as the pole was not stayed at all but was securely strutted instead with iron rods to the timbers of the roof. In another case where we employed a pad of tarred felt to check the vibration and consequent noise of a wire we were requested in future to use galvanised iron. Another amusing idea of an inspector was that four stays each capable of sustaining fifteen hundredweight were unsafe for a pole carrying one wire . . .

In 1894 the Company's agreement with Higgins was extended for a further five years. It will be remembered that back in 1878 he had stated that he would be content with £450 a year to cover his salary and any inventions he might make. This had been tactfully forgotten. He had already been given shares and frequent bonuses, and he put in for a rise of salary quite as often as anybody else. He always got it in the end. A chairman might not be indispensable, but the engineer was.

If Higgins was never quite satisfied with his remuneration

E

he was certainly not complacent about the instruments in his charge. He was forever making technical improvements and finding ways of cutting down expense. One such economy was effected by changing from batteries to dynamos and accumulators, first in the Cornhill office and then at the Haymarket. By the summer of 1891 all the general news circuits were worked by dynamo-charged accumulators with, said the Directors, 'economical and satisfactory results'.

But there were other things which were not satisfactory at all. The greatest single disaster was the action in 1894 of the Committee for General Purposes of the Stock Exchange in forbidding the supply of Stock Exchange prices to brokers other than members of the Stock Exchange. Fifty-five subscribers, using eighty-nine instruments, were objected to. The loss of these clients meant a loss of £3,466 revenue. This came in the middle of a period of business depression which had already hit the Company and was to continue until 1896 when the Exchange Telegraph was again in the red, to the extent of £648. At an early stage of the depression Kenneth Anderson's stockbroking firm had lent £3,200 to the Company, and the preference shareholders had been asked to advance 'the sum of £2 per share, being amount of the balance now uncalled for upon the A shares, such sum to be treated as a loan, carrying interest at five per cent. per annum.'

There were other causes than the general state of trade for the Company's difficulties. Chief of these was the bitter and often ruthless fight carried on between rival agencies, from which nobody profited except the lawyers. A number of cases were brought to court, but two will be sufficient to show their character and the expensive difficulty of controlling such behaviour by legal means as opposed to ordinary good faith.

Gregory and Co. had been subscribers to the Exchange Telegraph financial service, but the connection had ceased and their instrument had been removed. They managed, however, to induce a current subscriber to pass on the Stock Exchange prices he received, and Gregory and Co. then sold the information to other people. Their defence was that there can be no copyright of news, provided that it is not passed on in the same words in which the information has been previously embodied. But the Exchange Telegraph's contract for the supply of financial news (as for all forms of news) expressly bound the recipient not to pass it on. The court granted an injunction restraining Gregory and Co. from infringing the Exchange Telegraph's copyright and from continuing to induce any subscriber of the Company to supply them with information in breach of his contract.

The case got into the textbooks as one in which it was difficult to say how far the decision was based on infringement of copyright and how far on breach of confidence.

The second case was against the Central News and their associate the Column Printing Syndicate, to restrain them from copying information on the results of races at Manchester, issued by the Exchange Telegraph, and passing on this information to their own subscribers. They did not deny using the information thus obtained but submitted that the result of a horse race was public property. Mr. Justice Sterling, however, ruled as follows:

> The information was not made known to the whole world; it was, no doubt, known to a large number of persons, but a great many more were ignorant of it. By the expenditure of labour and money the plaintiffs [Exchange Telegraph] had acquired this information, and it was, in their hands, valuable property in this sense – that persons to whom it was not known were willing to pay, and did pay, money to acquire it . . . I think that it is established as against the Syndicate that they published for their own benefit information acquired from or through some subscribers of the plaintiffs, with knowledge or notice on the part of their Managers that it was acquired contrary to the terms imposed on the plaintiffs' subscribers.

An injunction was granted with costs. This sounds both straightforward and logical, but the case had been hanging over for months, being twice adjourned. More important, it did not put a stop to news-stealing. This could be done comparatively simply and in a moment, whereas the remedy of the law was slow and liable to be expensive. One comes on a note about 'the setting of another trap'. It was difficult to prove that correct news had been stolen. If false news was intentionally put out, and used by the thief, he went far to proving his guilt. But there was no direct remedy, for false news is fiction which is not covered by copyright.

There was generally a case of some sort pending or in progress. But this was not the most destructive activity of the rivals. More damage was done to both sides by price-cutting. The Annual Report for 1893 complained:

> The competition in the News Agency business referred to in the last report continues, and your Directors can only repeat their expressions of regret that the simple suggestion of agreed Tariffs has not been more favourably received.

The trouble was to continue well beyond the end of the century.

Another trouble – one which had existed from the beginning of horse race reporting – was the frequency of police raids on sporting clubs. This often resulted in the closing of a club, and in any case tended to scare off clubs which might have subscribed to the racing service.

The position in law was very briefly as follows. Under the Suppression of Betting-houses Act anyone who owned or used 'any office, house, room or other place' for the purpose of betting was liable to a fine of £100 or – if he failed to pay – to imprisonment for up to six months in the common gaol or house of correction, with or without hard labour.

This appeared to mean that there was nothing illegal in the Exchange Telegraph disseminating racing news. There was nothing illegal in a club receiving it. But they must not use the club as a place for betting, and few people are interested in horse racing unless they bet. More detailed interpretation depended on the definition of the word 'place'. A club was clearly a place; an enclosure on a racecourse might be, as the bar of a public house might very possibly be. But a man's private telephone was not. He could back his fancy by 'phone with legal impunity.

In 1894–5 two cases received a lot of publicity. Of the first there is a spendid account of the raid in *The Daily Telegraph* of 24 November 1894 and of the subsequent proceedings in the

*Evening Standard* on various dates in December, January and June 1895. They are worth summarizing. First the *Telegraph*:

Yesterday afternoon, shortly before two o'clock, the police entered the Albert Club, Bolt Court, Fleet Street and made one hundred and six arrests. The club is one of the principal sporting and betting resorts in the metropolis, and the news, which became widely known during the afternoon, produced a feeling of amazement, not only among betting men generally, but also in every sporting circle, the Albert being regarded as one of the best conducted, as well as one of the oldest-established, betting clubs in London . . .

The raid was effected in the most skilful manner. At the moment when the officers obtained an entry there was not a single additional constable to be seen in the vicinity, but within a few seconds every approach was guarded by several officers . . . Business was in full swing at the time, the Manchester November Meeting forming the principal subject of speculation. All the cash, betting books, and memoranda were at once seized, and the whole of those present at once placed under arrest.

Without any loss of time the whole of the prisoners were escorted in two's and three's to Bridewell Police-station [about one hundred and fifty yards from Bolt Court]. The majority elected to cover the distance in hansoms, rows of which at once formed on both sides of Fleet Street for a distance of two hundred or three hundred yards . . .

Several of the defendants were brought to the Justice Room at the Mansion House at half-past three o'clock, and the remainder followed in batches. The room was crowded with men, and room only could be found for the witnesses and the reporters. Many of the defendants were of good social position, and large sums of money were found in their possession . . .

The Lord Mayor took his seat at five minutes past five o'clock . . . Mr Crawford, the City Solicitor, prosecuted; and Mr Stanley, Solicitor, appeared for Mr Schrimshaw [secretary of the Albert] and the members of the club generally . . .

Mr Crawford, addressing the Lord Mayor, said that he appeared on behalf of the City Commissioner of Police . . . The intention of the Commissioner was to gauge the all-important point as to whether the club had been so carried on and the business transacted of such a nature as to be an infringement of the Betting Act . . . Doubtless legal questions would be raised hereafter on behalf of those who had managed the club, and it was desired that there should be the fullest enquiry, in order that a decision should be arrived at once and for all as to whether the club was legitimately within the law . . . Of course, if the defendants were able to satisfy his lordship that they had kept within the bounds of the law they would be honourably discharged . . . [An adjournment was then asked for.]

The Lord Mayor said that he would tell Mr Solicitor what he had decided to do. It was most inconvenient to have such a large number of visitors in the Justice Room of the Mansion House. (A laugh.) At the same time their convenience was the convenience of all who had to take part in the proceedings . . .

An adjournment was agreed, the defendants being granted £10 bail, except for Mr Schrimshaw, who was required to find surety for £100. He was an anxious man. His whole future was affected. The other defendants were not much concerned. The Lord Mayor had gone out of his way to be polite to them. The worst they expected was a fine of £100 – the equivalent of a misplaced bet.

The story is taken up by the *Evening Standard*. The case came up at the Mansion House on 5 December. The City Commissioner of Police had meanwhile decided to proceed against only fifty-three of the one hundred and six arrested:

The Lord ¡Mayor said that he did not view them in the light of ordinary criminals. (Applause.) They must not make that demonstration in court . . . He would not be able to take the case himself in the future, on account of his engagements . . . 6 December was then fixed as the day on which the case should be gone into fully.

On that date the case was heard by Alderman Sir George

Tyler. The hearing produced little except complaints against the behaviour of the police by counsel for the defendants, and defence of police action by Mr Crawford, the City Solicitor.

Mr Stanley and Mr Yelverton rose together, and a wrangle ensued as to who should speak, Mr Stanley insisting that since Mr Yelverton's client was discharged he had no right to speak.

Sir George Tyler called upon Mr Stanley, who observed ... the police took them [the Defendants] as they had a right to do, to the police station, and all they were entitled to do was to search them to see if there was anything on them relating to racing or betting; but what the police actually did was to take all their personal property – jewellery, rings, everything: it could not be suggested that a watch and chain or rings could relate to betting. Yet it was all seized. Then the police proceeded to question each Defendant, and to enter into a book whether they were married – surely it did not matter whether the Defendants were married or single if they were to be charged with frequenting a gaming house.

They were next measured –

Mr Crawford – No, no.

Defendants in court – Yes, yes.

Mr Stanley continued – The Defendants were measured in the ordinary way like a criminal. The colour of their hair was shouted out; the colour of their eyes – whether they were bright or otherwise. Surely these facts were not wanted in a charge of frequenting a gaming house. The system was a very improper one by which gentlemen like the Defendants were treated like criminals – habitual criminals.

Mr Crawford – The operation of the Act is well known.

Mr Bonner thought it was an Act which would require to be repealed. In all events it was an old Act, and it was the first case in which a club of known repute had been treated in that way.

Mr Crawford – No, no. There are several cases.

Mr Bonner – This was a well known club of social repute ...

Mr Yelverton wished to say –

Mr Stanley – He does not represent any Defendant today.

Mr Yelverton – I am surprised at Mr Stanley interrupting . . .

Sir George Tyler fixed the further hearing of the charges
for 13 December.

On that date the case came up before Mr Alderman Newton.
(No one magistrate presided over more than one hearing.) But
neither of the prosecuting counsel was present, they being
engaged in other important cases. The defending counsel were
absent for a similar reason. So the case was adjourned.

It came up before Mr Alderman Ritchie on 8 January with
Mr Matthews and Mr Muir for the prosecution and Messrs
Gill, Avory and Biron for the defendants.

George Powell, the informer, gave evidence:

> He attended the club regularly from June 1889 until the
> day after the raid, when he resigned (laughter) . . . No
> member was allowed to stand by the tape. There was a brass
> rail round it, within which a servant stood, who called out
> the names of the horses. When the billiard-marker called
> out the names of the horses Mark Myers, a member of the
> club, called out the odds against the horses. He had a clerk
> as partner named Lu Laden who wrote down the bets.
>
> Mr Matthews – A penciller, I suppose.
>
> Mr Avory objected to Mr Matthews imparting his racing
> knowledge into the case (laughter) or polishing the sentences
> of the witness (loud laughter).
>
> Witness – . . . The billiard-table when the marker was
> engaged was covered over.
>
> Mr Matthews – And what went on upon it?
>
> Witness – The members bet on it.
>
> Mr Avory – What were the odds against the billiard-
> table? (laughter.)

The case was continued on three other dates, and is described in many more columns. There was more wit than legal
argument. Finally the City Solicitor asked leave to withdraw all
the charges, and the case was dismissed. It cannot be said that
justice was seen to be done. But the result was approved by
most people – except the Anti-Gambling League.

The Anti-Gambling League took out summonses against the stewards of the Jockey Club 'for opening certain enclosures or rings on Newmarket–heath for the purpose of betting and for permitting such enclosures to be used by other persons for the purpose of betting on 10th and 24th October 1894, the days on which Cesarewitch and Cambridgeshire races were run'. *The Times* devoted half a dozen columns to the case.

The stewards were the Earl of March, the Earl of Ellesmere and Lord Rendlesham. Of the six magistrates two, including the Chairman, were clergymen. Mr Poland Q.C., who had failed to appear in the Albert Club case through pressure of other business, led for the prosecution. His opening speech lasted for an hour and fifty minutes, and the case occupied the whole of two days. The legal arguments centred on whether an enclosure on a race course was a place within the meaning of the Act, whether the enclosures were located as they were for the sake of betting or for watching the racing, and whether the stewards permitted such betting. It was contended for the defence that an enclosure was not in this sense a place. If any person exhibited 'any colour, clog, hatband, stool, umbrella for the purpose of betting' – in other words if any person carried on business as a bookmaker with the trappings of his trade – the spot on which he stood would be a place. But bookmakers were not permitted within the enclosures, although they might speak over the railings to the people within. It was further contended that the stewards did not knowingly or wilfully permit betting within the enclosures. The magistrate accepted these points and the case was dismissed. The Press account concluded:

> The decision . . . was received with applause in court and with much cheering outside. Mr John Hawke, the Secretary of the Anti-Gambling League, was the subject of a somewhat hostile demonstration as he walked to the railway station.

In this case, as in that of the Albert Club, it was stated by the prosecution at the start that their main purpose was to make plain *once and for all* what was legal and what was not. This was not achieved. The prosecution of small clubs continued, so that many were loth to accept the Exchange Telegraph's racing service which at once made them suspect by the authorities.

During this period there were a number of innovations. The Haymarket office, which housed the editorial, was kept open day and night under an editor named Boon. Perhaps in honour of this, electric light was installed instead of gas.

In 1891 a Law Courts service had been instituted. The Exchange Telegraph had this particular field to itself for a long time. But when other agencies followed suit the Courts became a field for battles no less ruthless than those fought on the race courses. When a civil case of great public interest was being tried, the reporter of a rival agency installed a private line telephone in a lavatory so that he could phone the result to his office within seconds of the verdict being given. The Exchange Telegraph's correspondent took along a colleague whose task was the sedentary one of locking himself in the lavatory!

In 1895 yachting was added to the service for newspapers. A paragraph of the letter to editors announcing this is worth quoting:

> This year's Yachting promises to be more interesting than that of any preceding year. The contests in which the *Britannia* and the *Ailsa* take part are certain to be of an exciting character. Mr Herreshoff, the famous American Yacht builder, will be represented by two new twenty raters, one built for Prince Leopold of Hohenzollern, and the other for Mr Howard Gould, while the 40's will include the German Emperor's new Yacht, designed by Watson, another boat by the same designer, and one by Fife. As regards the big cutters, interest will be mainly centred in the contests in which the *Ailsa* and Lord Dunraven's new *Valkyrie* take part, as one or other of these boats will be the challenger for the America Cup.

The foreign service had always been expensive. Some Directors had doubted if it was worth maintaining, but it had been kept up largely for prestige. Toward the end of the century it was increased by the appointment of extra correspondents. In 1897 a Mr Damala was given the assignment of covering the Turko-Greek War at a remuneration of 10s a message. Arrangements were also made for a local journalist named

Temple to telegraph Indian Frontier news. Correspondents were appointed in Copenhagen and Rome. Then came the South African war. Not only was a special correspondent appointed in Ladysmith, but one of the staff, McDonnell, was given six months' leave to join the army. Six months was then the maximum time that the war was expected to last.

The reporting of Test Match cricket in Australia was begun in the antipodean season 1897–8. Australia won the series 4–1. There was some sparkling cricket with Ranjitsinhji in the side. Two English and three Australian batsmen had averages of over fifty, and runs came fast in those days.

On the commercial side, financial advertising was started in 1899. There is a note to the effect that a nett profit of £308 was made on advertising the Victoria Deep Leads prospectus. But this new departure seems to have been on a small scale.

In this year the estimated revenue exceeded £50,000 for the first time. The break-down in round figures was: financial £20,500; general £15,000; provincial and sundries £14,500. But expenditure was running at £49,000. The Company continued to plough back all it could and was never far from the red line.

There were a number of changes in the Board. In 1891 J. H. Hutchinson died, and the Hon. Edward George Villiers Stanley was elected in his place. He remained a Director for only three years, retiring – as Lord Stanley – in 1894. The Hon. Schromberg Kerr McDonnell filled the vacancy.

In 1898 death struck with a heavy hand. At a special Board Meeting on 1 February the Managing Director reported the death of Lord Sackville Cecil, who had been Chairman for nine years, and also of another member of the Board, General Sir Michael Kennedy. Colonel Francis Sheppee and A. D. Goslett filled the vacancies, Colonel Sheppee being shortly afterwards elected Chairman.

Less than five months later another Special Board Meeting was convened. At this the Secretary reported that the Managing Director had died on 21 June.

The following was sent out on the Company's general news column printer:

The Exchange Telegraph Company deeply regrets to announce the death of its Managing Director – Captain Davies – which occurred last night. Captain Davies had been Managing Director of the company since its inception in 1872 and with Sir J. Anderson and Cyrus Field was one of its founders. To his great business ability was mainly due the prominent position which the Exchange Telegraph Company has assumed among the great news agencies. Prior to his connection with the Exchange Telegraph Company, Captain Davies had an eventful career in the service of the Pacific Steam Navigation Company and the P & O Steam Navigation and it was while acting as agent for the latter company at Aden that he rendered a very important service to the Government of the day in coaling the ships of the Abyssinian Expedition [led by General Lord Robert Napier in 1868] and received in recognition a service of plate from the Lords of the Admiralty. Captain Davies was chief officer on board the steamship *Great Eastern* on her maiden voyage to New York. Captain Davies brought into commercial life the best qualities of the sailor. A man of scrupulous fairness, outside his family circle he will be missed by none more keenly than members of the staff of the Company from the highest to the lowest.

Practically all the papers used this, some adding details. His death was due to influenza. He was remarkable for his energy and geniality. He was also on the Board of the Arcadia Coal Company of New York, the District Messengers Service and News Company (in which the Exchange Telegraph held one thousand shares), the Northfleet Coal and Ballast Company and the Swansea Improvements and Tramway Company. He was a Swansea man. On visits to his home town he used to tell his friends of his experiences at Aden, reputedly the hottest place in the world. What he there wanted most was to get out of his skin. None the less he created a garden, the only one in Aden, and the Arabs used to stare for hours at a time at his patch of flowers and greenery. He also had a cistern installed in his house, and into this he used to get, clothes and all. In consequence he suffered from rheumatism for the rest of his life.

Moustached, bearded, side-whiskered, and with a short neck, he looked pugnacious. But the lively, kindly eyes contradicted this impression.

There is a quite different point about this news item. It contains a mistake in its second sentence. One might expect that the American co-founder of the Exchange Telegraph should have been Cyrus Field, from the fact that he was so deeply involved in the Atlantic cable and also was a close friend of Sir James Anderson. But in fact he had nothing to do with the Company. In the Articles of Association the founders are given as Sir James Anderson and George Baker Field, landed proprietor, 131 Sixtieth Street, New York.

G. B. Field also attended the first three Board Meetings before he went back to America, never to be mentioned again in the minutes. He cannot have been even a relative of Cyrus, for Samuel Carter, the careful and excellent author of *Cyrus Field, Man of Two Worlds,* who had access to all the family papers, can find no record of him. It is interesting how a mistake can be made in a moment of stress, be accepted uncritically, and perpetuated.

At the special Board Meeting when the death of Captain Davies was announced, Wilfred King was promoted from Secretary to Managing Director. He was then thirty-seven. Probably many of the characteristics mentioned by pensioners and present senior members of the staff developed later. But enough can be said now to give at least an impression of W. K. Physically he bore a striking resemblance to the Prince of Wales, the future King Edward VII. This went far beyond the fact that he had an ample, well-trimmed beard and a large head (his top hats had to be specially made for him). The one might easily have been mistaken for the other. In private life he was ascetic, a bachelor, a non-smoker and non-drinker. He dressed meticulously in the fashion of the day. And he was a demon for work.

At the time of his appointment he gave one hint as to his character. It had been the task of the Secretary to write the minutes of Board Meetings. George Morgan was appointed

Secretary when Wilfred King was promoted. But W. K. held on to the quite arduous duty of keeping the minutes. One might perhaps see in this the tendency of a man who likes to hold all the reins in his own hands.

# CHAPTER 5

# The Survival of the Fittest

THE COMPANY celebrated the year 1900 by installing a large cast-iron lamp over the Pope's Head Alley doorway of the office in Cornhill, permission to do so having been obtained from the Public Health Department of the City of London. This lamp is remembered by some of the staff, for it was removed with the Company's headquarters to Cannon Street. There, extinguished for the blackout, it survived the blitz. But its late-Victorian style was unsuitable for the up-to-the-minute East Harding Street office, and it was 'disposed of'.

The new century started with a snowstorm as bad as or worse than those of the winter of 1888–9. When Higgins first reported on the damage two hundred and thirty subscribers were cut off, there were hundreds of miles of wire to be inspected, and lines were still falling. A fortnight later the number of interruptions had risen to three hundred and thirteen, and this before other line-owners, more dilatory than the dynamic Higgins, had started to pull their broken wires over the Company's mended ones – with destructive results. Conditions were so bad all over the country that the Post Office was reduced to sending telegrams to the north of England by train.

Higgins started work first on the Stock Exchange lines, with a staff increased to emergency numbers. But it was a matter of weeks before he had sorted out the mess. One may imagine a tangled skein of wires sprawled over the roof-tops. It was difficult for any workman, particularly if new to the job, to recognize the wire he was responsible for, and to repair it without further damaging the others. Besides, the different gangs were not particularly friendly. Days and nights on the tiles led to arguments.

Such a state of affairs made it all the more urgent to get lines underground. For more than ten years every effort had been made to do this, with memorials addressed to the Commissioners of Sewers. But London's underworld was already crowded almost to the limit, and there was plenty of red tape to snag the wires still further. The files contain requests by the Company for permission to lay lines under certain pavements, and an acknowledgement by the Commissioners who said that the matter would be considered.

At a lower level a first-class row ensued when an Exchange Telegraph gang met one of the City of London Electric Lighting Company in Fleet Street. The Company's men were accused of reckless behaviour, wanton damage and the extinguishing of arc lights under which they were working. Higgins is on record as replying that he was not one of those who wanted to take up the whole b . . . pavement. As a point of social history the word beginning with 'b' was never at that time written in full, for Fred Higgins preceded Bernard Shaw's Professor Higgins by a dozen years.

The chief engineer at this time of stress and strain applied to the Board for permission to apprentice his own son. This was granted. There is no hint in later records of any criticism of the boy. But neither is there praise. One would have thought he found it difficult, as every son of a great man does, to live up to father's reputation. But that he was not lacking in confidence and self-opinion is shown by his behaviour on his father's death.

The most lively subject of public interest at the beginning of the twentieth century was the continuance of the South African war – the 'Transvaal War' as the Board continued to call it although it had long spread far beyond the Transvaal. Queen Victoria's long reign, now ending, had been an era of peace similar to that under the classical *pax Romana*. Apart from the Crimea and certain trouble with tribesmen there had been no war. When some section of the Empire was provocative, an expedition was sent out and settled things almost from one issue of the newspapers to the next. Britain ruled not only the waves but a large proportion of the dry land as well. She was invincible. And here was this trouble with the Boers dragging on. We were losing lives, battles, prestige. The public was outraged, urgently awaiting the next news. The Exchange Telegraph appointed another war correspondent. There is also a nice domestic touch. The name of McDonnell, who had been granted six months' leave to join the army, came up before the Board after that period. He was given a six months' extension. There is no record of McDonnell winning battle honours, but since the war lasted for over two years he achieved the distinction of being the junior member of the staff most talked about at Board Meetings.

One of the Exchange Telegraph's correspondents did receive a silver medal. The Company put a great deal into the reporting of the war, and received more than £10,000 a year directly from subscribers for this service. Other agencies were no less active. The War Office, from which all official despatches were issued, was then in Pall Mall, on the site where the Royal Automobile Club now is. The Exchange Telegraph, with its editorial office in the Haymarket, appeared well placed. But the Press Association and the Central News managed to set up private telephone lines to their offices from a point just across Pall Mall. Thereupon the Exchange Telegraph obtained permission from the Junior Carlton Club, which was directly opposite the War Office, to install a private phone in the hall porter's box. Thus official bulletins could be copied immediately they were posted. They were not only sent out by the Company over the wires but also published in a broadsheet entitled *Telegraphic News,* price threepence.

British forces had been hemmed in at Ladysmith, Kimberley and Mafeking by November 1899. It was the resistance of Mafeking that most gripped the public imagination, for here the defenders were for the most part civilians who had been mobilized by the Commander of the small military garrison, Colonel Baden-Powell. They successfully defended their very vulnerable town for six and a half months. Meanwhile there was very little news, except about the progress of the force which was marching to relieve the town. When at last the siege was raised the Exchange Telegraph did not profit from its strategic outpost opposite the War Office, for the War Office, not yet having received an official despatch from Lord Roberts,

F

displayed only the notice 'NO NEWS'. It was a reporter from Reuters, the only agency with which the Company had not quarrelled, who first wired back the information which set London 'mafficking' for twenty-four hours and more.

But the Company was first in reporting the relief of Lady-smith after a siege of one hundred and twenty-three days. The newspapers reported as follows:

> The Ex Tel Coy says the following has been posted at the War Office from Sir Redvers Buller to the Secretary of State for War. Littletons headquarters 1 March 9.5 a.m. Dundonald with Natal Carbineers and composite regt entered Ladysmith last night. The country between me and Ladysmith is reported clear.

The South African War was followed by that between Russia and Japan. The Exchange Telegraph sent a correspondent out even before hostilities began. His dispatches were considered highly satisfactory. In using one of them the *Express* commented, 'Fighting must be a trying business at present, for the temperature on the Shaho is thirteen deg below zero'. (From this one deduces it was still considered that wars should be comfortable.) In September 1905 the Exchange Telegraph was first in reporting an armistice.

For the Company and the other agencies the immediately subsequent wars were concerned with the turf. The first began as skirmishes over the collection and transmission of news by course reporters and touts of doubtful loyalty temporarily employed. There were saboteurs, agents and double agents – all the ingredients of a secret-service story. And at managerial level there was a great deal of money at stake. The Exchange Telegraph, whose exclusive interest had originally been financial intelligence, was by this time fully committed to reporting sport which, with the minor exceptions of cricket and yachting, meant horse racing. Football and other ball games were not yet sufficiently organized to create mass interest. Horse racing was, in an important sense, a financial matter. Betting, though harassed by the law, and officially ignored as if non-existent by the Jockey Club, was even more of a popular passion then than now. So speed and accuracy in reporting results, and odds, were all-important. The Exchange Telegraph had adopted the motto 'Accuracy, Impartiality, Celerity'. It took a lot of living up to.

One factor bound to cause trouble was that the racing authorities would not permit telephoning from the course, believing that if punters could learn the result of a race within minutes of the finish they would not spend money to attend the meeting (a similar argument to current television coverage of football). Consequently, unless the number board could be read from outside the course, some sort of visual signalling was employed from the ground to someone who was beside a telephone, or – if the distance was too great – to an intermediary who could see, perhaps through a telescope or binoculars, both the tic-tac man on the course and also the man by the telephone to whom he relayed the signal. Celerity was by this means achieved, but accuracy was less certain as also was impartiality, for not only the Company's reporters were concerned. They were dependent on the course officials who posted the results, and who also might grant facilities to one reporter rather than another. (One must add in parenthesis that telephones were then very few and widely separated.)

These points were made evident in 1904 in the affair of Mr Ford of the firm of W. J. Ford and Sons who were responsible for the number boards at several race courses.

At Derby there was no telephone on the course. The nearest was at the Colwick Hall Hotel. The Exchange Telegraph arranged with the hotel-keeper to allow their reporters the exclusive use of this phone. From close to the hotel one of the number boards (not the judge's) was legible. Thus within seconds of the number of the winning horse being hoisted it was possible to phone the result of a race to the London office.

'Impartiality' meant to the Company fairness; it did not, of course, preclude competition with other agencies: that was an unwritten rule well understood by all reporters. But Mr Ford, for whatever reason, had a different interpretation of the word. He wrote to the hotel-keeper, whose name was Barron: 'I was very much surprised at your allowing the telephone to be used at the last Race Meeting and must ask you to see that it does not occur again'. Mild enough in writing: but Ford told

Barron that the brewers who owned the hotel, and presumably also catered for race meetings, would sack him unless he did as he was told. Barron, however, emboldened by a personal assurance from Wilfred King, continued to honour his agreement with the Company during the meeting of 24 and 25 October.

The next move was that the solicitors of the race course owners wrote to the Company's solicitors a letter which concluded:

> If they [the race course owners] are unable to prevent the use of the telephone at Colwick Hall for this purpose, notice will be given to the Telephone Company to remove their posts and wires from our client's land, which they are obliged to do upon receiving three months' notice. This will put an end to any further discussion, and we can see no advantage in continuing this correspondence.

This pithy statement may have been interpreted as bluff by the General Manager who knew all the tricks of the sport of kings. At any rate William Gibbs, the Company's chief racing reporter, who at the time of writing is ninety-four years old, was instructed to continue phoning from the Colwick Hall Hotel.

Ford then used other tactics. As has been said, the Company's spotter could only read one number board. His view of it was at an oblique angle. Ford fixed up a plank which cut off this view.

The Company thereupon started negotiations with the Derby County Cricket Club whose ground was within the periphery of the race course. From their pavilion there was a direct view of the board in question. And there was a telephone in the pavilion.

Ford must have got wind of this move. He anticipated it by hoisting false numbers on the board read by the Company's reporter, although the numbers on the judges' board were correct. He employed these tactics also at meetings at Nottingham and Leicester.

The direct result was that the winners of several races were wrongly published. Indirectly it was demonstrated how widely the Exchange Telegraph's service was used, for the writer has seen thirty-two papers which published the errors, and there were no doubt more. Newspapers do not like being put in the wrong, and they protested strongly when they published the corrections. One instance is given in brief, the other in full.

*Leicester Evening News,* 17 November 1904.

The hoisting of the wrong numbers at Leicester on Monday has created considerable discussion. Numerous complaints have been made by disgusted backers of Barbecue and Ballerine colt, who destroyed their tickets when they saw the numbers of March Flower and Vita on the frame as winners of the Town Selling Plate and November Nursery respectively. Now for such mistakes there is no excuse. March Flower and Vita did not finish in the first three, and moreover the correct numbers were given by the judge . . .

*Bristol Daily Mercury,* 21 November 1904.

As everyone connected with racing knows, there are a large number of sharpers ever ready to prey upon the public and their tricks are legion. The introduction of telegraphy gave them a fine opening, but the telephone was their trump card, and many an honest bookmaker has been robbed by its aid. The evening papers have of late discounted that *modus operandi* by issuing extra racing editions, and one of the foremost in this respect is the *Bristol Echo,* whose editions with the results of races are sold in the streets within five minutes of the result of almost every race.

The Exchange Telegraph Company have made every effort to disseminate this news; but last week their efforts were frustrated, to a great extent, by the stupidity (?) of the officials at Leicester and Derby, as on no fewer than six occasions was the wrong number hoisted.

Surely this ought to have been inquired into by the acting stewards of the meetings referred to, and ought to be brought before the stewards of the Jockey Club, as not only does it cause endless bother to punters on the course, but is a loophole for possible sharp practices, and, goodness knows, there is too much scope for that already.

The Exchange Telegraph themselves brought the matter before the stewards of the Jockey Club by means of a printed letter of four and a half foolscap pages, dated 30 November, plus a number of appendices which included affidavits by the principal witnesses. It was a very strong letter, hinting in one paragraph at something like high treason:

> The Directors much regret to say that the above false information was consequently supplied over the Company's instruments to His Majesty the King at Buckingham Palace, and to H.R.H. the Prince of Wales at Marlborough House.

On 7 January following, the Secretary of the Jockey Club wrote a personal letter to Wilfred King saying that the stewards had considered the document and had just expressed to Mr Ford their 'very strong disapproval of his actions in the matter.'

This disapproval did not, however, prove strong enough to eradicate such practices.

In February 1905 the sporting pages of the newspapers all carried articles on the Starting Price War – or S. P. War as it was unaffectionately called. Until that time *The Sportsman* and *Sporting Life* had each had their trusted representative in the ring at race meetings, and together they had agreed upon the list of starting prices which were thereafter sent out by the agencies and published by the newspapers. Then the two sporting papers disagreed, and began sending out separate S. P.s which sometimes differed widely.

The *Weekly Dispatch* reported:

> The result is chaos . . . And the remedy is not easy to find. The Jockey Club is barred by its own statutes, and by the politic reason which supports the particular rule immediately concerned, from recognizing betting, so no remedy can be expected from that quarter . . . The present situation is an absurdity. It is rendered possible by the existing relation between betting and English law.

English law did not help, the next Betting Inducements Bill promoted being entitled 'An Act to prevent the writing, publishing, or circulating in the United Kingdom of advertisements of any betting or tipster's business.' In 1912 – for the trouble went on as long as that – the Exchange Telegraph held a plebiscite among bookmakers to discover which followed the *Sportsman's* figures, and which those of *Sporting Life*. The former proved the more popular. But neither side won the S. P. war. It rumbled on until comment on it was drowned in the thunder of the First World War. Today, starting prices are fixed in much the same way as they originally were.

With the law antagonistic – although little enforced by magistrates against public sentiment – and the Jockey Club sitting blindfolded on its fence, not recognizing that anybody ever backed a horse, it was inevitable that there were other squabbles like that with W. J. Ford and Sons. But it must not be supposed that the Exchange Telegraph spent all its time quarrelling with members of the horse-racing fraternity. It quarrelled with a good many other people too. It frequently went to court against those who stole its news. There was, for instance, the case of Frederick Howard, Managing Director of the Manchester Press Agency. Howard was suspected – more than suspected: he was known to have tapped the wires – of using the Company's cricket news for distribution to his own customers. To catch him, a score was given one or two runs wrong – a neat trap, for the result of the match was not affected. Howard fell into it, three times. In court Mr Justice Buckley gave judgement against him, not for breach of copyright but for fraud. *The Times*, which in both measurement and prestige was a still bigger paper than it is now, devoted four of its solid-set columns to a report of the case – a measure of public interest.

By far the fiercest quarrels were with other agencies, principally the Press Association and to a lesser extent the Central News. Struggle for survival would be a better phrase, for that is what it was with the P.A. at least.

The Press Association had been registered in November 1868. It started work as soon as the Post Office took over the private telegraph companies in 1870. It was launched as a co-operative association to collect and supply news to any provincial paper which wanted telegraphic news. Only provincial newspaper proprietors could become shareholders. The Exchange Telegraph Company never had any newspaper

men on its Board, and its original object was as stated in
Chapter Two. The Association and the Company were as
different as they could be. Consequently they were friendly – at
first – for although human couples have a better hope of
concord if they have interests in common, the opposite is the
case with news agencies.

The struggle which developed in the first years of this
century is briefly and clearly described by George Scott in his
*Reporter Anonymous*, the centenary history of the Press Association. His account is of particular interest, for if the author were
biased at all it should be on the side of the P.A. In fact he is
commendably objective – and one is tempted to quote him at
length. But a somewhat fuller account will be attempted, for
the affair was of major importance, involving the first real
challenge to the existence of the Company.

The Exchange Telegraph had begun to cover general news
in the early 1880s. It wanted to buy the P.A.'s service for its
subscribers in Liverpool and Glasgow, and the negotiations
only broke down over the price demanded. Later the Company
had begun collecting and distributing its own general news,
but at the turn of the century it was not yet capable of doing
this on a large enough scale to alarm the P.A. Apart from
finance and commerce, it was only formidable in the field of
sport.

On its side the P.A. had ventured into sport at about the
same time – 1880. Sport, as has been said, meant racing and
a little cricket, with interest in football only gradually developing. The P.A. was not at first geared to compete with the Company in this field.

To appreciate the finer points of the struggle one needs to
know how dependent the agencies were, or had been, on outside
people (most of whom they thoroughly disliked), and the
difficulties they faced in distributing the ever-growing quantity
of sporting news to an enlarging public against increasing
competition. Reuters, one must say at once, did not join the
battle. Concentrating on foreign news, they quarrelled with no
one – except perhaps Dalziel's agency which flourished in the
1890s and had the temerity to challenge them in their own
field. The P.A. had a friendly and mutually profitable arrange-

ment with Reuters. The Exchange Telegraph supplied Reuters with instruments. But the possession of a friend in common did nothing to improve relations between the P.A. and the Company.

The P.A. got their original racing service from C. H. Ashley's firm of Ashley and Smith, proprietors of *The Sportsman*. But they quarrelled with them in 1883 and worked with the Hultons of Manchester instead. Their relations with the Hultons are said to have been stormy at times, but by their quarrel with Ashley they made of him a formidable rival, for he covered cricket as well as racing. George Scott says that Edmund Robbins, manager of the Press Association, 'saw Ashley not only as a dangerous enemy but one whose existence was injurious to newspapers.' What Ashley thought of Robbins is clearly stated in a memorandum written by Wilfred King after a visit to *The Sportsman* office:

> He [Ashley] referred to Forbes and Moore of the Central News as unmitigated scoundrels, and Robbins of the Press Association as one of the most unprincipled men he had ever met . . . Mr Ashley then said that since his separation from the Press Association, Robbins had on many occasions stated it as his most earnest desire that he might before his death succeed in ruining Ashley and Smith, but Mr Ashley said so far he had not succeeded; but he hoped, however, that when Robbins died he would go to heaven.

Wilfred King enjoyed this confidence because the Exchange Telegraph was still working with Ashley and Smith. When the Company covered a race they gave the results to *The Sportsman*, and vice versa. Incidentally, the accounting must have been complex: there was constant bickering, even over quite small sums. But, comparatively speaking, the Company and *The Sportsman* were good friends.

What Edmund Robbins and Wilfred King thought of each other is not recorded. Of King's character we already know something. At the Cornhill office he had started from the lowest rung and had been pushed up rather than climbed, for there is never a hint of personal ambition. He was a bachelor. When he became sufficiently important to be included in

*Who's Who* the world read that he was Managing Director of the Exchange Telegraph Company – no more. He omitted date of birth, recreations, all the other details that people are invited to put in. One deduces that he saw himself as Managing Director of the Exchange Telegraph – no more.

Edmund Robbins had started work on a newspaper at the age of eleven. If ever there was a dyed in the wool journalist it was he. He had helped to keep the P.A. alive through the chaos which followed its birth and the nationalization of the telegraph companies. He had worked hard and successfully for the Association for more than thirty years since then. It was unlikely that he would give anything away to a rival. But he appears to have had a professional weakness. He was one of those rare and admirable people who dislike the telephone.

Wilfred King, brought up in the same nursery as Fred Higgins, looked on every mechanical thing as a slave of the Company. The Exchange Telegraph had bought a telephone for the office and one hundred for its subscribers as early as 1881. Certainly its reporters generally sent their copy by telegram until the turn of the century. But as soon as the facilities were available they reported by telephone, and the necessary office arrangements were made to receive news in this form.

The Company and the Association were engaged at every level, from the managerial to that of the most junior reporters. These latter suffered from no inhibitions. They were up to every trick. But even if a man succeeded in handing in his telegram first it might, for reasons beyond his control, be dispatched or delivered after that of his rival. In any case telephoning was quicker.

Not until July 1905 did the P.A. decide to launch a telephone service. It put the organization of this into the hands of Alan Greaves – whom, incidentally, the irascible Charles Ashley describes as 'an unmitigated blackguard.' Whatever his character, Greaves worked hard, fast and expensively. But a telephone service cannot be arranged over-night. There must, for instance, be copy-takers to receive the messages and, of course, sufficient instruments and lines. George Scott says that Edmund Robbins, still sceptical, told his board that 'the one way of keeping pace with Extel [this form of the name was not yet used by the Exchange Telegraph] in the telephoning of racing results and betting was to capture a trunk line from each race course. This meant booking a long series of consecutive calls and so holding open the line to London. The costs soared.. '

At this stage the Company withdrew the offer it had made to supply the P.A. with tape machines; and simultaneously it suggested that they should try to work together instead of against each other, since only the Post Office and the National Telephone Company were gaining from such a density of sporting coverage on both sides.

The P.A. turned down the suggested co-operation, and it ordered German tape machines. Higgins managed to get a look at these and reported that 'the instruments were not suited for multiple working and under the system of working adopted, to work an important number from one transmitter would be impossible.' As usual he was right.

The Exchange Telegraph had from the first set its face against rate-cutting. So had the P.A. A minute dated 27 September 1905 states that Robbins personally assured King that they would not cut their price. But in the heat of their struggle this boomerang was used. The Company, which for a number of years had been making a steady profit, showed a loss of £3,600 that year. The P.A.'s losses on the telephone service may have been twice as great. It began to look as if both the contestants might perish.

Then, in the early spring of 1906, the P.A. had the bright idea that all its troubles would melt away if it took over the Exchange Telegraph. Edmund Robbins and Wilfred King held a series of secret meetings at the Howard Hotel. One would give a lot for a record of what they said to each other. But all we have is the result. This was not a take-over but an agreement to work together. Wilfred King had evidently made an effective counter-attack.

The Treaty of Howard Hotel, which became known as the Joint Service Agreement, was signed on 3 July 1906. The purpose was frankly set out in the preamble:

Whereas for some time past the Association and the Company have been competing with one another in the collection of and distribution by means of telephones and typewriting telegraph instruments of Racing Cricket Football and other descriptions of news in the United Kingdom and such competition has been found to be detrimental to the interests of both the said parties and it has been agreed that with a view to putting an end to such competition and reducing the expense of the collection and distribution of the said news the said services (except and subject as hereinafter set forth) should be worked jointly and the expenses and receipts pooled. . .

So far so good – except for the parenthetic phrase which begins with 'except', and which could only lead to trouble. What it came to was that the provincial staffs should be fused, with consequent reductions, except at administrative level. Each would retain only half its telephone operators, half its reporters. But the Joint Service would be run by joint managers (Robbins and King) and supervised by a joint committee. All receipts would be pooled – except those of the Company in the county of London. (That was the main exception, but there were others.) Accounts would be kept jointly, and audited jointly. In fact the Association and the Company (as portrayed by the agreement) might be compared to Siamese twins who did not trust each other an inch. There was an arbitration clause which did not specify precisely the type of person who should arbitrate.

The joint managers worked together peacefully for a month or two. Then it was decided to send out a circular about coverage of the coming football season. But the partners could not agree as to whether they should both sign the same circular or each sign identical circulars individually. When the Joint Committee also failed to resolve the question the P.A. said that it must go to arbitration.

The minute recording the Company's reaction to this insistence suggests resignation to the inevitable:

The Board . . . decided that if the Press Association still insist on the question of joint circulars being submitted to an arbitrator, they will raise no objection to this course being pursued.

But it was not pursued for some time after that, for the two sides could not agree who the arbitrators should be.

Since so petty a disagreement could cause so much trouble it is not unnatural that there were many more. There was no attempt to conceal the mistrust which each side felt for the other; for instance the Joint Service accounts were audited as such and then audited by the Association's accountant. The Company's board meeting minutes are peppered with serious discussions about apparently unimportant points. It makes discouraging reading and is nowhere worth quoting. More important, these squabbles did not end with Boardroom argument. They were carried on before an independent arbitrator. But, as George Scott puts it, 'arbitration was all very well but not when it went against you'. So they were carried further, to the Law Courts, to the High Court. The shorthand notes of a High Court case in 1908, hingeing on the interpretation of the words 'all news' in the agreement, are as lengthy as a Russian novel.

The truth was that the Company and the Association had an agreement to work together but no will to work together. It took them the best part of five years to realize that what they were saving by their Joint Service they were spending on legal costs. Only then did they settle down to *make* it work. And in fact it did work very well.

It is more precious for having given rise to a recorded joke by Wilfred King. He said that the Joint Service should be called PAX.

# CHAPTER 6

# King's Reign Begins

NOTHING has been heard of Higgins for some time. He was far from idle. Early in 1907 he launched a 'new fast financial service' by providing instruments which worked twice as fast as those formerly used. This involved more than a thousand individual trials, 'which to insure uniformity as far as possible I conducted myself', he reported.

Possibly as a result of the consequent improvement the Stock Exchange asked for four more reporters, a striking change from the days when only one had been grudgingly admitted to the House. The three-man Stock Exchange sub-committee which dealt with the Exchange Telegraph visited the Cornhill office, were shown round and entertained.

Other changes initiated by the engineer were that the financial and general news instruments in the clubs were converted to the improved type, and there was a considerable increase in the length of cable laid underground.

Even more than most people, Higgins liked to be proved right. He generally was. His estimates of cost were nearly always very close to the actual figure (he never failed to point this out), and his assessment of new instruments was just about infallible. But in one instance he was far from the mark, possibly because the factors were more psychological than mechanical. He had prophesied that the House of Commons would never require more than three annunciators. In fact the number rose to fifteen. The *Western Express* gave the reason in a sceptical little article:

In some of the lounges, smoking, eating or drinking places of the vast Palace of Westminster there are little mechanical contrivances called annunciators. These machines are worked from a distance and throw up in large letters the name of the M.P. who is at the moment addressing the House.

A fresh name clicks out its letters on the annunciator. The M.P.s in the room look up at it with a languid glance, and resume the perusal of the Bluebook, the volume of Hansard, the letter from an irritating constituent, or the French novel. The name is that of a man outside the charmed circle of speakers who fill the House.

In another half hour or so comes a name that allures them from their comfortable armchairs. 'Balfour's up,' says the first man to read the announcement. They spring up from their upholstered depths, toss away the ends of their cigars or knock out the ashes of their pipes and hurry off to the Chamber in which most of their constituents fondly believe that they spend laborious days.

The annunciator became one of the characters of Westminster. When the newspapers were short of an inch or two they turned to it for a fill-up. The *Daily Mail* told this story about a new M.P. named Chancellor:

> On the occasion of his first speech a thin House suddenly became a full one, and almost as suddenly became thin again. Members, unfamiliar with his name at that time, had thought that 'Mr Chancellor' was short for 'Mr Chancellor of the Exchequer.'

The *Standard* included this in its question-time report:

> Mr Gulland (L. Dumfries) asked the First Commissioner of Works, in view of the advisability that members of the House should be aware of the proceedings in the other House whether he would place in the members' lobby an indicator showing the subject under discussion and the name of the speaker in the Lords.
>
> Mr Stanley Wilson. – In view of the legislation which is to be brought in to destroy the House of Lords, is not this question entirely unnecessary?

There had to be an Irish story. The Irish Party dinner of

1918 was held in the Strangers' dining-room. As the guests warmed up they became more and more annoyed with the English annunciator which remained preoccupied with the slow progress of the Lottery Bill in the Chamber.

Mr Hazelton tried tearing the tape, but that had no effect, and Mr Flavin, more drastic, stuffed a tablecloth into the mechanism. Even with this gag the machine clicked out the figures of the division, and refused to be silenced until the House rose and left it with its duty well done.

An annunciator was also installed at Euston to give information about trains.

Higgins was cool about the brain-child of another inventor which aroused considerable interest at this period. *The Times* of 5 December 1908 carried this report:

The Lord Mayor attended this week at the G.P.O. to transmit the first message to Manchester by the 'telewriter', a new process by which over an ordinary telephone wire a message is received in the actual handwriting of the sender. The Lord Mayor by this process conveyed to the Lord Mayor of Manchester cordial greetings to him and his fellow citizens from the City of London and himself. The reply was, 'Warmest greetings and congratulations. The Inventions of science unshackle the progress of mankind.'

The Telewriter Company suggested that the Exchange Telegraph should buy the instrument for the Stock Exchange service. The Directors were interested and remained so for some time, as various minutes show. But Higgins saw little future in the telewriter, and he seems to have been correct.

The engineer's passion for statistics has already been noticed. In his report for 1910 there was something to catch the eye. The number of instruments in subscribers' offices had risen to 1,100. The length of tape used was 19,000 miles, ample to reach round the world in the latitude of London. From a series of tests it was found that the average time interval between the announcement of a New York stock price and its publication by the Company's instruments was five and a half minutes. Enquiries for these instruments were by this time being received from as far afield as Russia, and a gentleman with the almost biblical name of Baltazar wrote in for details from Brazil.

Higgins had often had brushes with the authorities about overhead wires. In 1910 a by-law came into effect restricting the span to one hundred and fifteen yards. Higgins's comment was typical:

> The manner in which the figure of one hundred and fifteen yards was arrived at was rather curious. The London Chamber of Commerce requested my views on the proposed bye-law, and among other suggestions I recommended the span to be one hundred and fifty yards. When the matter was being discussed at a meeting of the Chamber, someone asked what the practice was on railways, and it was stated one twentieth of a mile span. The one hundred and fifty yards suggested plus eighty-eight yards of the Railways divided by two gave one hundred and nineteen yards. To give an appearance of decimality four yards were knocked off making one hundred and fifteen yards.

Higgins's linesmen were by the nature of their work more prone to accidents than were other members of the staff. In June 1907 two of them were recovering in the Swanage Cottage Hospital from falls, the Company paying the fees. One notices this, for the Board does not appear to have been over-generous in the early days to the employees or their dependants when they suffered damage or death. The widow of a linesman killed on duty generally received £5; and when W. Jeans died after being the House of Commons correspondent for twenty-five years his wife received a grant of fifty guineas. But on this occasion the Board discussed the question as a whole, and as a first move decided to give a subscription of three guineas a year to the Swanage Hospital. This was a beginning. (One must, of course, consider the value of these sums at the time. For instance, in 1908 the Company bought the football business of J. Bell and Co. for £25.) Another welfare gesture of the same year was that the Company offered to pay the fees of any member of the staff who wished to attend evening lectures, and they later guaranteed full pay for six weeks to any man absent through illness.

G

In 1909 the lease of the Haymarket office which housed the editorial was running out and could not be renewed. To avoid similar inconvenience in the future it was decided to purchase freehold, and Nos. 14/15 Panton Street, S.W.1. were bought for £11,000. It was paid for partly by selling stock, and partly by mortgage. From the accounts it looks as if the property could have been bought without borrowing, for the Company had been making a comfortable and steadily increasing profit every year since 1906. (This healthy trend continued until 1915.) Financially, the Joint Service was working well. It was going much more smoothly too. Counsel was still retained by both sides, and there were occasional cases for arbitration, but King and Robbins had come to prefer the cheaper practice of settling their differences between themselves. To look ahead a little, the Exchange Telegraph and the Press Association played a cricket match on 15 October 1913. It seems late for a fixture both in season and year – seven years after the agreement was signed. But this may surely be taken as a ceremonial burial of all hostilities. The staff, with old-world courtesy, wrote a letter to the Board, thanking them for arranging the game. There is no record of which side won. Perhaps it was a draw.

In this new spirit the Company and the P.A. worked together to catch out a news thief, R. J. G. Dutton, Manager of the Northern and Midland Reporting Agency. Before the Parliamentary Election they arranged that the *Launceston Weekly News* should subscribe to this agency. When the returns came in they purposely altered the figures in four cases – Hatton, Hoggerston, West Ham, Birmingham – taking care not to alter the majority. They subtracted a figure or added a figure to each side. These incorrect figures were telephoned by Dutton's agency to the *Launceston Weekly News*. This evidence was so convincing that Mr Dutton did not waste money on Counsel but conducted his own case. His defence was the ingenuous one that he had lifted the news from the Joint Service but did not know that it was copyright. Apart from having to pay costs, he suffered only to the extent that he had to accept a perpetual injunction against doing the same thing again.

Wilfred King visited the United States and Canada in the spring of 1911. He described the trip in a forty-four page report to the Chairman, some of which makes interesting reading. He called on the United Press Association which had been supplying the Exchange Telegraph with news, but not the right news:

> Mr Howard [the Vice-President] said that his difficulty in giving us a satisfactory service had been the difference between English and American ideas of news, that what they in New York would consider 'a spicey bit' we in England strongly objected to.

King was much impressed by the electric light advertisements in Broadway, particularly by that of the Heather Bloom Skirt Company which

> represents a young lady with her skirt blowing about, and the sign is certainly most attractive . . . the mechanical arrangements occasionally go wrong and an amusing instance of this occurred when the rain went the wrong way and ascended instead of descended.

He travelled far and held conversations with a remarkable number of people. But his two main purposes – canvassing American financial advertising and arranging for the setting up of an office in Canada to disseminate United Kingdom news – were unsuccessful, the first because of the keen local competition, the second from expense. 'Personally,' he concluded, 'the result of the trip can best be considered a bitter disappointment.'

In March 1913, Lt-Col F. F. Sheppee died. He had been the conscientious and reliable Chairman of the Company for fifteen years. The other members of the Board at this time were Kenneth Anderson, who had been a Director since 1886, G. A. D. Goslett, appointed in 1898, Horace L. Hotchkiss, the American Director who lived in New York, George Montagu, appointed in 1902, and Wilfred King who had been Managing Director since 1898. A senior and experienced Board. There is no record of the discussions on who should be Lt-Col Sheppee's successor. But it is unlikely that Kenneth Anderson would have accepted the Chairmanship, for he was still in active charge of a stockbroking firm which had more than once helped the Company by loans during the early difficult years. Horace Hotchkiss was not in the running. George Montagu had a mere eleven years' experience, and he seems to have had other interests, for he was not infrequently absent from Board Meetings. It is not surprising, therefore, that Wilfred King, then aged fifty-three, was nominated Chairman. More and more decisions had of later years been left to the discretion of the Managing Director, and one notices from the minutes that when he was away on a tour difficult questions were apt to be held over to the next meeting – and his return.

Typically, W. K. did not relinquish the post of Managing Director when he became Chairman. He was a big enough man for the double responsibility and we have already noted his inclination to keep as many reins as possible in his own hands. The Company was all the more surely guided. But the minutes of Board Meetings occasionally make one smile – very respectfully – as when reading that 'the Managing Director was instructed to see Mr Satterthwaite', and 'various applications for increases [of pay] were left for the Managing Director to deal with', signed Wilfred King, Chairman; and that the Managing Director's salary was increased by £300 [to £1,500]. As Chairman only he would have been worse off than many junior members of the staff, for at the forty-second Annual General Meeting – the first over which King presided – it was resolved that a Director's fee, be he Chairman or not, should be £200 plus a proportion of £100 which was to be divided among all the Directors at their own discretion.

Six months after Sheppee's death Kenneth Anderson died. He had joined the Board three years after the retirement of his father: that was the only gap so far in representation of the Anderson family. The two vacancies caused by death in 1913 were filled by E. C. Barker and Stanley Christopherson.

The first major task undertaken under Wilfred King's Chairmanship was a reduction of capital and additions to the Memorandum of Association. The Company had not been able to pay a dividend on the B shares since 1882, while the arrears on the A shares at the end of the last financial year amounted to £111,000. The scheme, very briefly, was to reduce the capital from £246,250 (8,125 A shares, 16,500 B shares, both categories of £10 each) to £96,430 (8,023 A shares of £10 each and 16,200 B shares of £1 each). This was done by cancelling forfeited A shares, by cancelling capital which had been lost or was 'unrepresented by available assets' to the extent of £9 per share on the issued B shares, and by cancelling unissued B shares.

The additions to the Memorandum of Association demonstrated the extent to which the Company had broadened its field of activities. That of 1872 mentioned only transmitting stocks and shares and business intelligence by 'wires and apparatus'. Now was added the collection and distribution of news of every kind in any part of the world. Advertising agency business was added. Financial advertising had been done for some years, nearly all the business being brought in by the Secretary, George Hamilton, 'from his personal friends'. They must have been many and rich, and he had done very well on a commission of twenty-five per cent. Now there was talk of an advertising department.

The objects of the Company would also include banking, co-operation with similar companies (such as the Press Association), to deal in shares, to 'borrow, raise or secure the payment of money in such manner as the Company shall think fit', to 'distribute any part of the property of the Company *in specie* among the members'.

There had been a good deal of argument before the Meeting, particularly about the reduction of capital; but on the day – 30 July 1913 – this motion was passed with only one dissenting vote, and the additions to the Memorandum of Association unanimously. About this time a canvass was started of cinemas, both in London and the provinces, to explore the possibility of providing them with news to be projected on their screens. Since this failed to provide convincing data it was decided to run a three months' service free of charge as an experiment. Presumably this was unsatisfactory, for there is no further mention of it.

At the beginning of 1914 conversations were held with the Secretary of Creed Bille & Co. *The Sportsman* had begun to use the Creed tape machines, and the Company had therefore become interested. They must have been at something of a disadvantage where such decisions were concerned, for none of the Directors had any technical experience of printing. But neither had they of electricity by which all their instruments were operated. They may have felt that they did not need this practical knowledge. After all, although none of them was a journalist they had successfully competed with the Press Association whose managers and members were all newspaper men. And in questions mechanical the Company always had Higgins on call.

It is perhaps surprising that Higgins was not made a Director. It may have been thought that he was sufficiently occupied as it was, and in a field where his genius was entirely beneficial – as it might not have been in the Boardroom, for among his numerous qualities diplomatic expression was not included. But the value of his service was recognized. His salary rose to £900 a year, second only to that of the Managing Director; and in February 1914 he was paid £1,500 for his transfer to the Company of his outstanding patents.

When during the winter of 1913–14 he was absent through sickness – for the first time in his long service – Wilfred King, a man who did not squander sympathy, corresponded with the Higgins family asking for news and insisting that the engineer must not think of returning to work until he was fully recovered from his 'chill'. It must have been in King's mind that the Company would be very much poorer if, to use the common euphemism, anything happened to Fred Higgins.

This chapter ends the description of an era. A summing-up is therefore appropriate, and that it may be from a contemporary point of view it is taken from a contemporary issue of *The Times*. A full page then appeared about the work of the Exchange Telegraph. Presumably it was an advertisement although, apart from the number of superlatives, it is written as a straight account.

Every journalist and most City and Club men are familiar with the tape machines of the Exchange Telegraph Company . . . Even the most hardened newspaper man finds himself at times admiring the exquisitely devised electrically-driven mechanism in the little glass case, and the brains behind it which have thus made it possible for the news of the whole civilized world to be collected, transmitted and delivered in visible form. It is not too much to assert that the wonderful organization of the Exchange Telegraph Ltd., and their still more wonderful transmitting and receiving machines constitute one of the most remarkable features of modern civilization. Every reader of the newspapers is indebted to a greater or less extent to the services of the Exchange Telegraph Company, and never more so than when, as lately, the morning and evening journals have been eagerly scanned to ascertain the latest news of the terrible calamity which is still fresh in the public mind. The 'Exchange' was the first in cabling over to this country the account given by the survivors of the ill-fated *Titanic*. . . .

The organization provides correspondents covering the whole world; while for reports of Parliamentary and legal proceedings as well as for Stock Exchange and Market prices, the Company has its responsible representatives in both Houses of Parliament, in the Courts, on the floor of the Stock Exchange, and in most of the London markets.

Similarly for obtaining first-hand expert accounts of Racing, Football and Cricket Fixtures it possesses a staff upon whose knowledge and skill it may depend for a prompt, accurate and graphic account of the game or race. How minute the arrangements, and expeditious the transmission of news may be better understood by following in some detail the actual process of obtaining and transmitting to the Press (and to private subscribers) a description of some well-known Turf event in which a large share of the public is interested, and to announce which the evening papers will issue a special edition.

In the case of a great race, say, at Epsom, the Company has one representative on the Grand Stand in communication with another representative who has a telephone office near the winning post and telephones direct into the London offices of the Exchange Telegraph Company. Preparations such as these are, of course, made in advance, and it is often necessary to carry a wire directly onto the scene of action, and this is done at all football and cricket matches. Before the starting of the horses a brief preliminary account is sent over the lines descriptive of the scene. The moment the race is started the fact is communicated from reporter No. 1 to reporter No. 2, and thus within the space of one minute the news that the horses are 'off' is known not only in London, but at Manchester, Liverpool, Glasgow, and all the great centres of population in the country. Similarly, as soon as the winner is declared, the announcement is known in the Provinces within a minute of it being known on the Grand Stand – often, indeed, before many on the course have been able to ascertain the order in which the horses are placed . . .

The Company has prepared a table showing the average duration of the classic races, and is able therefore to check very closely the work of its representatives in collecting and transmitting the news . . . A delay of a few seconds will be reported to Headquarters, and will possibly pass without an enquiry into its causes, but should a message arrive at one of the Exchange Telegraph Company's Provincial offices considerably – say a minute – behind the regulation time, an investigation into the cause of delay will be made from Head Office. One of the results flowing from this marvellously rapid communication of racing news is that it has practically stamped out the fraudulent practice which was possible when an unscrupulous punter, having received a private telegram or telephone message as to the result of a race, betted on a certainty.

The celerity of the Company's operations is not limited to the sporting news service, and the arrangements made for the report of Mr Winston Churchill's recent speeches in Belfast and Glasgow well illustrate the elaborate similar arrangements and with similar promptitude they despatched some 3,000 words of the speech delivered in that City by the Rt Hon gentleman.

The Exchange Telegraph Company offers seven distinct instrumental services at varying annual charges. Thus, for fifty guineas per annum, any newspaper or private subscriber may obtain the Parliamentary Service. This means that a machine, printing the news on tape or in column form, will be installed at a convenient point on the subscriber's premises, and that during the Session a summary of the proceedings of both Houses of Parliament will be delivered by the instrument, extending when the occasion requires to about ten thousand words for each sitting. The report is continuous, being despatched over the Company's special wire from the Reporters' Gallery, and records the proceedings within a few minutes of their occurrence in the House. In this connection one may perhaps mention one of the most remarkable of the Exchange Telegraph Company's instruments, due, like the majority of their machines, to the inventive genius of Mr F. Higgins . . . One refers of course to the Annunciators . . .

The General News Service involves the use of two machines, which between them supply Metropolitan, Provincial and Foreign news to the extent of 11,000 to 16,000 words per diem . . .

The Legal Service comes direct from the Royal Courts of Justice in the Strand by means of a fast printing instrument – a staff of reporters specially engaged by the Company covering the whole of the Common Law, Chancery, Divorce, Probate and Bankruptcy Courts, and telegraphing their 'copy' as the cases proceed. Verdicts and important incidents are recorded within a minute after they are known in the Courts. The system of delivering large batches of copy by hand has been practically superseded by this service, the instrumental delivery being continuous, and enabling sub-editors to bring the report of a case down to the latest moment for the various editions of their papers . . . And very extensive arrangements [are made] for the report of an enquiry held in some building other than the Courts of Law – the enquiry into the loss of the *Titanic* being a case in point . . .

For the sixth service – the Financial – [the Company] possesses pre-eminent and exclusive facilities for the collection and distribution of Financial Stock and Share reports, the trustworthiness of which is assured by many years of practical experience. The Company is the only News Agency telegraphically connected with the Stock Exchange and allowed to have Reporters and Telegraph Clerks in the House for the collection and despatch of Stock Exchange quotations. In this respect therefore it possesses unique advantages.

In addition to the above named facilities . . . the following London Markets are at present being reported – Billingsgate

Fish, Borough Hop, Central Meat, Borough Potato, Corn, Produce, Provision, Coal, Metropolitan Cattle, Whitechapel Hay and Straw, London Metals and Tallow Sales. In addition to these, reports are also given of Glasgow Pig Iron Market, Bank Return, Freight Reports, Floating Grain Cargoes, Eastern Exchanges, Silver Market, New York Stock Exchange prices, American produce, Bankers, Clearing House return, and Bullion movements.

The tariff rates for the vaious services above named range from about fifty guineas to one hundred and fifty guineas per annum . . .

When figures get into millions the mind refuses to comprehend their magnitude. For the curious however it may be stated that the total number of words transmitted by the News Services of the Exchange Telegraph Company for the year ending 30 June, 1911, amounted to no less than 607,065,598 – a figure which represents an annual output of printed material equivalent to 6,000 good sized novels.

Quite apart from the seven distinct instrumental services above detailed, the Exchange Telegraph Company undertakes at any time the temporary supply of news of any character to any person at any place in the United Kingdom or abroad – wherever in fact telegraphic communication is possible.

This statement of activities appeared soon after the Company's fortieth birthday. It had moved a very long way since the days when it was buying a few American instruments and cautiously negotiating with the Stock Exchange for permission for one of its employees to stand upon the sacred Floor. Now it was very proud of itself in public, though still highly critical in private, and at all times confident of the future – although putting a thousand or two pounds in reserve whenever it could, just in case.

# CHAPTER 7

# *War*

IN THE financial year which ended on 30 June 1913, the Company made a profit, in round figures, of £8,200, in 1914 of £9,500, in 1915 of £16,600. In 1916 the profit was down to £1,700, and in 1917 there was a loss of £400.

The reason for this change of fortune is evident enough – the World War. What is remarkable is that the Company went on making money for so long when services were drastically curtailed and few newspapers or clubs could afford to pay for them at the existing rates. Clearly a great effort was made by the Directors and those of the staff who did not put on khaki (none joined the navy). We will try to follow their fortunes.

The business of the first wartime Board Meeting began exactly as every other had – minutes of last Meeting; financial memoranda (amount invested £13,000, amount on deposit £3,000, Pool account £2 16s 3d overdrawn, etc.); Estimated rate of revenue £92,500. Only when these items had been dealt with did the Directors turn to war news.

Circulars which had already been sent out were examined. (Clearly the Chairman and Managing Director had acted on his own.) Their subjects are not recorded but they may well have concerned the restrictions on categories of news – movements of battleships and troops, for instance – which had been made shortly before hostilities began. Major Stuart of the War Office had enquired whether the Company wished to send a correspondent to the front: the Board cautiously resolved to see the regulations before deciding. *The Times* and *Daily Mail* (then both owned by Lord Northcliffe) asked whether the Company would allow its war news to be used on their bulletin boards: this was turned down. The *Daily Mail* offered their war news by telephone at 10s 6d a week or by telegraph at 2s a message; this modest offer does not appear to have been accepted. The Company was evidently determined to maintain its self-sufficiency.

Then the Managing Director reported that five Army Reserve men, all operators, had been called up and that five Territorials, operators and mechanics, were already with their battalions. It was resolved that their pay should be made up to the peacetime figure or that they should be put on half pay – whichever gave the greater total – and that their berths should be kept open.

The Board then went on to discuss Club, Stock Exchange, and Press rate telegraphic services, subscriptions (decision deferred), and bookmakers' subscriptions which they agreed to cut by fifty per cent if necessary. Clearly and naturally they were uncertain how such things would be affected.

The Directors can scarcely have been reassured by the behaviour of the Editor, John Boon. On 14 August he and Wilfred King had a row, and as a result of his 'conduct and behaviour on that occasion the Managing Director had been compelled to suspend him until the next meeting of the Board'. Boon did not wait for its verdict. Only four days later the Managing Director read in the *Daily Mail* an account of the battle of Dinant, the first engagement between French and German forces on Belgian soil. The date-line was Brussels, 17 August, 'From our Special Correspondent, John Boon', *The Times* carried an identical report 'From our Special Correspondent'. The Board decided to dismiss John Boon – rather like locking out the horse after it has been stolen – and went on to discuss the appointment of a war correspondent.

Within two months of the beginning of hostilities twenty-seven members of the staff had joined the army, and the Company had decided to take out war risk policies for the Cornhill and Panton Street buildings. The former was insured for £6,000 and the latter for £10,500. The total premiums were £20 13s 6d which suggests that the underwriters did not consider the danger great. In fact the first damage was suffered in October 1915 from a Zeppelin raid, but it was only to the overhead wires; and shortly before this a Government insurance policy against aircraft damage direct or indirect had also been taken out for Cornhill and Panton Street, and for Creechurch Lane, the store and workshop, and Whitefriars Street to which the engineer's office had moved when the lease of the Fleet Street premises above the Green Dragon pub expired. The Board were convinced insurers, covering every risk they could foresee. Thus when the Brighton manager embezzled £100 they claimed and received this sum in full from the Commercial Union Assurance. Thereupon the Board insured for the same amount the man with whom they had replaced the defalcator.

With 1915 the difficulties of the Company became more apparent. On 4 January the Stock Exchange reopened for the first time since the outbreak of the war, but business was still subject to 'considerable restrictions,' the publication of buying and selling prices being forbidden. As for the general news, the Censor's blue pencil greatly reduced its appeal.

The Managers of the Stock Exchange granted the Company a rebate on the rentals of call rooms for the quarter during which they had been closed, but many of the subscribers asked the Company for reductions both of the last quarter's subscription and those to come. These were granted at between twenty-five per cent and fifty per cent 'with the object of keeping as many subscribers as possible on the Company's books'. In May the Jockey Club cancelled all racing for the season, except at Newmarket. Irish racing continued, but it was more expensive to cover (trunk telephone charges had gone up thirty-three and one-third per cent), and the bookmakers were less inclined to pay for the service even at a reduced rate. Subscriptions for the special sporting service, the cricket and the football service had to be cut by a quarter to keep names on the books. Telegrams were also much more expensive. The Chancellor of the Exchequer had raised the Press rate by two hundred and fifty per cent. The clubs intimated that the subscriptions they had been paying for the various services were now more than they could afford. Higgins's annual report in July showed reductions all round, notably in the number of Stock Exchange calls which fell by 1,600,000 to a mere 457,000 ... So it went on, service costs rising and subscribers expecting rebates.

How were these difficulties met? First by economies, although economy had always been a first principle since Wilfred King had become Managing Director during a period when the Company's finances were at a low ebb. The Directors cut their own salaries by half. This was a gesture, for the savings amounted to only a few hundred pounds, but it set the tone. Against this the lower-paid members of the staff were finding it increasingly difficult to manage on their wages with the cost of living going up fast. They were given more than usually frequent rises – only of a few shillings a week but enough to encourage them. They also received certain bonuses, in 1915 to a total of £355. They must have worked hard, for men continued to join the army, and replacements were almost impossible to find. One reads of a discharged soldier being engaged, but does not know why he had been discharged. He may have been crippled. As for the men who joined up after the first six months of war, the Board decided that bachelors could no longer be given an allowance unless the circumstances were exceptional. Apparently they always were exceptional, for there is no record of any recruit not receiving an allowance, if only of 5s a week.

This scarcely sounds like economy, nor does the fact that all new investments were in war stock while most of the existing holdings were sold to buy it, often at a loss. But one specific way by which the Company's business was kept going was that the Managing Director made numerous and extensive tours throughout the United Kingdom to talk with and maintain the loyalty of subscribers, even to find new ones. During the month of December 1914 alone, he visited York, Middlesbrough, Newcastle, Sunderland, South Shields, Edinburgh, Glasgow, Newcastle again, Birmingham, Liverpool, Manchester, Stockport. He maintained this sort of pressure throughout the war when travelling was difficult. A number of manuscript letters and memoranda which have remained suggest how he kept pace with his regular office work meanwhile, and the Chairman always presided at Board Meetings.

In a search for more money various innovations were made. One was the war guinea service. Subscribers received important news by telegram to that amount. This was evidently a success, for a guinea service remained when peace returned. The Company ventured into the literary field when in conjunction with *Land and Water* they syndicated daily comments by Hilaire Belloc to a number of provincial papers at five guineas a week. More significant financially, it persuaded a number of its subscribers almost to double the subscription they paid for column printers. But not until the end of 1917 did the P.A. agree 'under protest' to pay more for theirs. One may comment that, while the Company's instruments still led the field, a lot of money might have been earned by making them for sale instead of only for hire to subscribers; but this might not have been possible during the war owing to the shortage of materials.

In February 1915 a French reporter, André Glarner, was earmarked for the Western Front, and the correspondent in Russia was instructed to get as near to the fighting as possible. Later, a man named Moseley was sent to the Dardanelles.

The first war casualty among members of the staff was reported in May when C. D. Copinger was killed in action in France. But by far the most serious loss to the Company occurred on 1 September. Frederick Higgins died. There had been little warning. He had been ill and absent from the office in May but had soon returned, first for half days and then full time. The observant might have noticed that at the General Meeting on 28 August he failed to propose that the auditors should be re-elected. In compliance with the Articles of Association, Deloitte, Plender, Griffiths & Co. retired every year, and from the very beginning Higgins had proposed their re-election. It seemed a point of unwritten law. But in 1915 he was, perhaps, too slow on his feet. Someone else proposed the motion and he only seconded it.

Four days later he was dead. His funeral at Abney Park Cemetery was attended by Wilfred King and forty members of the staff – a large proportion of the personnel of Head Office and the London Engineering Department in their depleted form. 'The Directors wish to place on record,' it was minuted, 'their sense of the serious loss which the Company has suffered by the death of its Engineer, and their great appreciation of the services rendered by him to the Company during a period of over 43 years' – since 1873 when he was twenty-two.

'Serious loss' was an understatement. The Exchange Telegraph would not have achieved the success it had, and might indeed have gone under, if instead of Higgins it had possessed an ordinary engineer.

Higgins came to the Company from the Island of Mauritius where he had been Superintendent of Telegraphs – a very young one. The only instruments which the Company then possessed were the Edison double-wheel tape machines which required one wheel to print the wording and another the figures, and which transmitted six words in a minute. Higgins at once started making improvements, and by 1902 had produced a single-wire instrument which transmitted at thirty-five to forty-five words a minute – six or seven times as fast.

He made a number of other inventions, too, during this period. In 1874 he produced an electric call device which enabled subscribers to communicate their requirements to the Exchange, and, under the same patent, the first fire alarm system ever to be placed on the streets of London. This he greatly improved in 1882. It was in the pre-telephone days that the possibility of calling the Exchange was of the greatest value. When the telephone came, Higgins promptly improved this on the loud-speaking side. But he was discouraged from continuing work on it by the monopoly which was given to the United Telephone Company, later called the National Telephone Company.

In 1880 he produced his column printer, 'a boon and a blessing to newspaper men,' as the Press described it. This wrote the matter in columnar lines on a strip of paper about as wide as a newspaper column, as opposed to the tape machine which ticked out horizontally a line-wide 'tape' of paper. He felt that he had perfected the tape machine by 1902, but the column printer he went on working at for the rest of his life. In 1889 he made an instrument which could print in either tape or column form, working from one transmitter on a single circuit. In 1887 he had begun using accumulators for power – before the Post Office or anybody else did so. For the technically interested this was described as consisting of 210 L 15 cells, the capacity being of approximately three hundred ampere hours at the ten-hour rating.

And there was, of course, the annunciator which has several

times been mentioned but has not been technically described. It printed a line three or four feet long in capital letters just over one and a quarter inches high. In the transmitter the type was hand-set in a long narrow carriage and moved below electrical contacts which communicated corresponding impressions in a mosaic form to a slowly passing band of paper at the various receiving points. The wording transmitted was calculated to fit exactly the framework through which the paper passed at the receiving points.

All these inventions, with innumerable minor improvements, were to the profit of the Exchange Telegraph.

There was another way in which Higgins served the Company: he publicized it in the most effective way – by making it talked about at home and overseas. The name Higgins, though difficult to pronounce across the Channel, became well known abroad. As early as 1893 the *Revue Internationale de l'Electricité* carried a long article headed 'Telegraphe Imprimeur Higgins'. After technical description of his numerous inventions this concluded:

> L'usage courant que la *Telegraph Exchange Company* fait en Angleterre de ces appareils ingénieux est un sûr garant de leur bon fonctionnement, et on ne peut qu'applaudir à cette importation chèz nous d'un instrument dont les avantages seront sûrement appréciés par toutes les personnes qui ont intérêt a être rapidement informées.

*La Nature* also devoted an illustrated article to his work – although it might seem to belong less to nature than to art.

Higgins's caustic humour must have been apparent from some of the quotations from his reports. Of his loyalty and humanity there can be no doubt. He surely worked while others slept, to achieve all he did. His inventions were the fruit of talent. But his constant testing, his ubiquitous interest, showed devotion. He pounced on any mechanical thing which might be to the disadvantage of the Exchange Telegraph – the copying of an instrument by a competitor, for instance. His monthly reports, packed with statistics, must have taken many hours to compose. He always gave full credit to the gangs who worked for him, backing them up if they barged into trouble.

H

Unfortunately there is no record of what education he received, but he had the crystal-clear brain of a scientist. Gordon Dain who occupied the same position of chief engineer forty years later – a very long time in proportion to the historical span of electrical engineering – found that Higgins's brain had achieved something which is now considered impossible without modern equipment. He says:

Around 1956-7 it became obvious that a review should be undertaken of the means of transmission and reception of the 'ticker tape' signals, due to the introduction of new telegraph relays and the coming general application of transistors. Experiments were carried out, using modern analysis aids such as oscilloscopes, to determine the optimum conditions for transmission. From these experiments a mathematical analysis of the line and instrument conditions was developed which not only aided future circuit design but also gave an explanation of the success of the late F. W. Higgins's work in the ticker tape field. It may be added that, without the use of modern test instruments mentioned previously and the progress made in the telegraph field in the past half century, such a mathematical analysis would have proved much more difficult and, in F. W. Higgins's days, would have needed a person of great creative intellect.

The brilliance of Higgins may now be judged, as a few years later – during the move of Extel Head Office to Extel House – some of the original papers of Higgins were unearthed, including one which described in words the mathematical analysis which took place so many years later. He had worked this out from first principles.

Although Higgins must have put out hundreds of thousands of words of manuscript reports, and although a large number of journals printed articles on him and his inventions, he himself only had one very small book published, and that privately. This was an *Inspector's Note Book,* pocket size, printed in 1881. Only one copy of this can now be found. It is in the possession of a member of the engineering department who was persuaded to lend this rare thing to the writer. One would like to reprint it complete, but it is for the most part too technical for general reading. A few extracts will convey the foresight, thoroughness and individuality of the author.

## INSTRUCTIONS

The following instructions are issued for the general guidance of Inspectors in charge of the Company's instruments.

Circumstances will occasionally occur for which no hard and fast rule can be provided, and in such cases much must be left to the judgment and intelligence of the Inspector.

Inspectors must bear in mind that the chief feature of the Company's business is instrumental delivery of news, and that however valuable such news may be to subscribers, if the instruments are not kept in proper order to perform the work required of them the service becomes valueless.

Attention should always be paid to the wishes of subscribers, but it is desirable that communications intended for the Company be obtained in writing.

Verbal instructions to remove or shift instruments should never be accepted.

Civility to subscribers and their clients is imperative.

## RULES

This Rule Book must be always carried by the Inspector into whose custody it is given, and returned when he leaves the service.

Each Inspector will upon entering the Company's service be required to make himself conversant with the instructions contained herein, and upon leaving the service must render up all the Company's property in good condition, the effects of wear excepted.

## DUTIES

1   Each instrument is to be thoroughly dusted and oiled at least once a fortnight.

2   The connections are to be examined each time the glass shade is removed in order to ascertain that they are secure.

3    No instrument is to be cut out of circuit or removed from an office without authority, and instruments must be changed without interrupting the circuit or causing loss of intelligence.

4    An instrument is not to be allowed to work without paper . . .

10    In the event of the bad working being caused by something being out of order in the instrument itself, and which it is out of the Inspector's power to discover or remedy, the defective instrument must be immediately changed for another . . .

## FAULTS

13    Do not trust to Providence for the removal of faults.

14    Any defect in the wires outside houses, or any building or other operations which threaten to interfere with the wires, are to be promptly reported . . . [The rest of 'Faults' is highly technical].

32    The Inspector's tool bag should contain –
One large straight screw-driver.
One large curved screw-driver.
One small straight screw-driver.
One hook for replacing spiral springs.
One pair of cutting pliers.
One bottle of oil.
One bottle of instrument ink.
Some spare spiral springs ready for use.
One tooth brush for cleaning type wheels.
One clean duster.
One piece small wire three feet long (and in locker ready for immediate use, about twenty-four feet of small wire).

## POCKET BOOK

33    A pocket book for the entry of complaints, instrument changes, and circuit lists is to be always carried and used.

34    Inspectors must be careful to keep always within reach of signals which may be sent upon their instruments . . .

42    Specimens of printing are to be brought in from each instrument once a week, and after notice to clean or adjust a

particular instrument a specimen should accompany the reply which announces that the cleaning, etc., has been effected.

F. HIGGINS
Engineer

Higgins's inventive brain was never idle. He was fond of the office cat. He used to feed it. But he was not always in the office. No one was there on Sunday. One cannot usefully give an animal a double ration, for it merely overeats at one meal. So Higgins devised an instrument which dispensed the correct amount of cat meat at fixed intervals. The cat became tuned in to the machine and visited it at the appropriate time.

Higgins's son, Clifford, refused to serve under his successor, E. G. Tillyer. It seems that he expected to be promoted to his father's place, which it was felt he was not competent to fill. The result was that he resigned, and the connection of the Higgins family with the Exchange Telegraph was severed.

One cannot help feeling sorry for Tillyer, thoroughly efficient and Higgins-trained though he was. Apart from the disadvantage of following a genius he had more than his fair share of troubles. There was a destructive snowstorm during his first winter. Shortages due to the war were becoming progressively acute, and within five months of his appointment John Gregory, the Company's Superintendent, died after forty-two years service. Tillyer must have felt lonely.

His first annual report, in July 1916, gives some idea of his difficulties:

No useful comparison can be drawn, the Stock Exchange having been closed for six months in the previous year. Owing to the war there were no telegrams received from foreign Bourses. 49,395 announcements have been placed on the tape of the names of Stock, and telephone numbers of Brokers wishing to deal, against 22,335 last year. In this case also no useful comparison can be drawn . . . seven hundred and thirty-one errors were made by the Reporters, an increase of one hundred and nine on the past year . . . Materials we use in construction and maintenance have considerably increased [in cost], metals averaging fifty per cent., wire ninety-three per cent., ink fifty per cent., and battery materials one hundred per cent. to five hundred per cent., and in addition we have experienced difficulties and delay in obtaining them as in nearly every case a permit had to be obtained from the Minister of Supplies before the manufacturers would entertain an order.

On two occasions our lines were damaged by Zeppelin bombs. On 8 September in Liverpool and 13 October in the Strand, the latter being serious owing to a seven-wire cable (overhead) being severed near the Lyceum theatre practically cutting off all communication with the newspapers, costing for repair £9 16s. We have erected an additional wire between Charing Cross and Carmelite Street . . .

Numerous faults have developed during the year in different sections of our underground system, caused by premature failure and decay of the gutta percha covering. This matter has been thoroughly investigated in conjunction with the makers and it is admitted the quality of the present day gutta percha is inferior to that obtained twenty-five years ago.

With regard to staff, Tillyer reported 'We have been able to maintain the curtailed services so far with the help of the willing hands still remaining'. But thirty-two of his men had joined the army. Although he had found some replacements he was still short of five inspectors, eight operators, six mechanics, three linesmen, one battery man, one clerk, and two inkers. By the next year the position was worse. There were no linesmen at all to remedy faults in Liverpool, Manchester and Leeds; and 'female operators have been introduced'.

Ladies were also being employed at Head Office. They had to wear all-concealing overalls of dark blue, and according to tradition (it is not minuted) the Chairman and Managing Director insisted that they must be plain. Anyone who sees the pretty girls who now decorate the office feels very sorry for their predecessors.

At the beginning of 1918 the staff were given a second and larger bonus – 5s a week for married men, of whom there were fifty-three, and 2s 6d a week for the seventy-four bachelors. This cost the Company £1,125 a year. Judging by private correspondence with other Directors, Wilfred King was very keen to push this through. Shortly afterwards, on representations by foreman printer Shrimpton to the Managing Director, all the printers received a further small increase. Although the last annual account showed a loss, the Company was evidently at pains to keep the good will of its hard-working staff.

In October 1918 The Exchange Telegraph had the distinction of being the subject of a column-length article on the main news page of *The Times*. This reported the libel action brought by the Prime Minister against the Company. Lloyd George had left London on the evening of 25 September directly after it became known that enemy aircraft were approaching. The E.T.C. – as the Company was then generally called – sent out the news flash:

> The Prime Minister spent the night at his residence at Walton Heath, Surrey, having left Downing Street about the time that it first became known that the raiders were approaching London.

This went to eight newspapers, of which three published it – the *Star*, the *Westminster Gazette* and the *Western Mail*. That was more than enough. Lloyd George had actually left London for a long-standing arrangement to meet the French Prime Minister at Boulogne. Certainly he had heard of approaching enemy aircraft just before he left. He had found Charing Cross in darkness and his special train had been held up throughout the raid on the bridge across the Thames. He had seen the shells bursting and heard the bombs. He had gone on to Dover where he slept before crossing the Channel.

Lloyd George insisted on going into the witness-box. Having explained what actually happened, he said that as a private individual he would have let the libel pass. If he fought every one who attacked him he would have no time for anything else (laughter). But the suggestion that the Prime Minister had left London on receipt of advance information that an air raid was imminent was likely to undermine the confidence of the munition workers of the East End. However, in view of the unqualified apology of the Exchange Telegraph he waived damages and only asked for costs. There is no record of what happened to the reporter or sub-editor who sent out the message. He ought to have been promoted for the publicity that he inspired.

There was another non-war casualty in this year. Norman Herbert, who had been first the only E.T.C. reporter at the Stock Exchange, and latterly chief reporter, died after forty-two years' service with the Company. Higgins, Gregory, Herbert – the first crop was dying off. But they had created a tradition. There are still men on the staff with similar length of service. In this case the family connection was not broken, for Herbert's son, Ralph, was in the office.

In 1917 there came trouble of another sort. Its sudden birth is described in a memorandum of the General Manager dated 20 March:

Visit [from] Mr Bradbury, Secretary of Lloyds Bank to say that he had come to give me, he feared, a shock. I asked him what it was about, and he replied that they had decided to give us notice to clear out of these offices [15-16 Cornhill] and their Solicitors would be preparing a formal document which we would probably receive in the course of a few days. He said they were very sorry to give us the notice, but they found they were constantly wanting more room and it was apparently due to the fact that they had to take on more and more women to do the work and had always to provide additional cloakrooms, lavatories &c. for them; that some little time ago he had been told by his Board that he would have to provide two hundred women: he had done so, and now the figure was approaching two hundred and fifty. As far as he could see, the women had come to stay, and although he had personally fought against their introduction into the business as much as possible, and had been the last man to take on a girl clerk, he was now compelled to admit that they had proved quite satisfactory, and if he was asked whether he wanted his male shorthand clerk back, he thought he would say no and would prefer to go on with the ladies who, he was bound to admit, did their work most excellently and seemed to take greater interest in it than the men.

This caused an urgent hunt for other premises, and an appeal (not ignored) to Lloyds Bank for a longer breathing-space; for it was evident that even after new premises were found it would take time to redirect all the lines. There was finally a short list of two buildings, one in Crown Court, Cheapside, and the other in Cannon Street. All the Directors went to look at the two premises, except possibly E. C. Barker who was working at the War Office. At the beginning of November 1918 an offer of £33,000 was made for the freehold of 64 Cannon Street.

Ten days later the guns stopped firing.

CHAPTER 8

# *Jubilee*

THE COMPANY faced the everlasting peace expected by the optimists with 1,116 instruments in subscribers' offices. Satisfactory as that was, it gives an idea of the amount of work involved in moving equipment and redirecting all the lines from Cornhill to the new office. The cost of this and of alterations to the new premises was little short of £5,000. The purchase price of 64 Cannon Street, of £33,000, was paid partly by mortgage and partly by selling investments.

In a peace-inspired spirit of goodwill the Press Association and Central News agreed to co-operate with the Company in reporting the General Election of December 1918 – there is not a hint of friction. (This was a most important election, for it was the first to be held since 1910). Members of the staff still with the Forces were given leave to help in the collection and transmission of the results. The Managers of the Stock Exchange – where H. Rutherford had taken over as chief reporter – waived entrance fees and subscriptions for that year.

The National Union of Journalists reacted differently. They negotiated for a minimum wage higher than the maximum which the Company was paying its sub-editors and reporters. The Company offered 5 guineas. But they had to fall in line with the other agencies, and the figure of 8 guineas a week was finally agreed. This was fairly generous at the time, and remained as much as most people got during the lean 1930s. But it was not revised for a quarter of a century, by which time the price of living had risen considerably. The National Union of Press Telegraphists followed the journalists' lead, but they asked and received no more than £4.

Everyone got rises in salary during the immediate post-war years. Wilfred King began granting small ones to individual members of the staff on his own initiative, only telling the Board about it afterwards. The Board invariably approved. But these were never large, amounting only to a total of a pound or two a week spread over several applicants. The recipients always wrote to say 'thank you'. In May 1919, the last bonus which the staff as a whole had received was made a permanent addition to their wages. One or two foreign correspondents were granted another £100 a year. The auditors asked for and received higher remuneration – £600 a year for the monthly bookkeeping and £50 for the yearly audit. A pleasant little touch is that Tillyer inherited from Higgins the unwritten right to propose the re-election of the auditors at the Annual Meeting.

The Board did not forget themselves. A Director's fee rose to £300, with an extra £100 for the Chairman. The Managing Director was excluded from this, but for the managing half of his duties Wilfred King received £2,000 a year plus ten per cent of the Company's profits above £7,500. He did not always claim this. In 1920 the profit was £12,500, and it continued to rise.

The case of George Hamilton also deserves individual mention. He asked for £1,000 a year, and was given it with the information that it was the maximum that a Secretary of the Company could expect – although his salary in fact rose to £1,200. But this was less than half what he actually earned. The advertising agency department was flourishing and he continued to receive twenty-five per cent on new business, which generally meant a three-figure cheque every month. The three or four other members of this department were entitled to the same percentage, but earned much less.

The subscribers were not pleased with the inflationary tendency, for they were asked to pay the pre-war rate, or more. But very few gave up the service. In general, the number of subscribers rose.

A department even younger than the advertising agency was the statistics service. It started in 1919, taking over from the Roneo Company a duplicated service of financial information. The service was printed by Messrs J. J. Keliher on cards which the subscriber filed in a cabinet which fitted them, supplied by the Company. There was a separate card for each British Company of sufficient importance, and when any announcement was made about one of these companies the appropriate card was reprinted and reissued to all subscribers. In effect the statistical service not only supplied its subscribers with information but did their filing for them.

It was launched behind a barrage of circulars, which produced sixty subscribers straight off. Then S. Forde Ridley who headed a staff of not more than half a dozen (including R. W. Brash, who later became manager, and A. Collett who became assistant manager of 'stats') began travelling round the country canvassing. Every fortnightly Board Meeting was told of an increase in the number of subscribers. Before the end of May (the service had started in April) there were one hundred and twenty, by June one hundred and sixty-one, and the figure continued to grow. A report on the statistics service, compiled in 1943, states that no accounts are available for the first eighteen months – January 1919 to 30 June 1920. That is so. But a search of other sources reveals that the four hundred and eighteen subscribers of February 1920 represented £7,755 4s 0d – a comfortable income. In December the department was strong enough for the Company's Board to suggest to the Roberts Statistical Company 'that perhaps the two companies might work in co-operation to the extent of each agreeing to keep to its own particular line of work'. The founder of the rival firm, Major Douglas Roberts, died without answering the letter which made this proposition.

One gets a sense of tremendous vitality and activity from all the notes one can dig up about the Company at this period. Even when the change of premises was imminent the Managing Director went off on a tour – to Aberdeen, Dundee, Edinburgh, Glasgow, Newcastle, Sunderland, South Shields, Middlesbrough, York, Hull and Scarborough within a fortnight; then after two days at head office he was off to Bristol, Birmingham and Coventry. And on the same day the Secretary left for Ireland.

Tillyer, though not a genius, was as active as Higgins had been. And like Higgins he suffered from the English weather which celebrated all important occasions in its own way. A violent snowstorm had ushered in the new century. There was another in the winter when Tillyer took over. Yet another storm welcomed the peace, or perhaps the change of premises, for it occurred on 27 April 1919, breaking forty-seven circuits and silencing three hundred and fifty-five instruments. The Cannon Street office had been occupied less than a month previously – on 30 March – involving much extra work for the engineers. The peace terms were promulgated on the night of 7 May. The editorial staff were up all night, writing a 5,000–word summary and a further 1,500 words of 'Points,' which were 'much appreciated by the Newspapers'. But all this could only have been transmitted if the lines had been repaired.

Tillyer comes out as a trustworthy and sympathetic man. In his reports he always praised his staff who were still working under the disadvantage of wartime shortages of materials and skilled men. And the words he used strike one as sincere. He was enterprising too. In 1914 Higgins had made trials in long-distance transmissions, working between Hull and Grimsby. These experiments had been interrupted by the war and halted by his death. In 1920 his trained disciple, who must have assisted him six years before, picked up the strands. On Sunday 3 October experiments were carried out from Cannon Street to the Company's Nottingham office and back, via the Central Telegraph Office of the G.P.O., a distance of about two hundred and seventy-five miles. In the first experiment a Post Office neutral B relay was installed at the end of the return leg of the Nottingham line, and through this relay a fifteen-ohm receiver was worked using a local battery of ten volts each side, positive and negative. The transmitter at Cannon Street was revolving at a speed of one hundred and ten r.p.m. The voltage to line was varied from sixty-six to seventy-six, with a steady current of fifteen m.a. and a running current of ten m.a. Results were very satisfactory.

Three weeks later a similar experiment was carried out to Newcastle, using the *Newcastle Chronicle* wire. The Post Office neutral B relay, fifteen-ohm receiver and local battery were in Newcastle. Transmissions from Cannon Street were carried out through primary, secondary and tertiary relays. Reception at Newcastle was very good at speeds of one hundred and ten to one hundred and fifteen r.p.m. but not satisfactory at one hundred and twenty r.p.m.

A further experiment was made on 9 November. The circuit this time ran from Cannon Street via the Central Telegraph Office to Bristol and back to Cannon Street, with both transmitting and receiving units in the Cannon Street office. After seventy-five minutes the transmission was transferred and continued from the Parliamentary Service at the Panton Street office. The transmitter of the Parliamentary service was rotating at one hundred and twenty r.p.m. Reception was satisfactory.

The Post Office neutral B relay which was used is now in the museum of apparatus in Extel House.

Although it means stepping outside the period covered by this chapter, it is tidy to record here that as a result of these experiments the first service to subscribers from a relay office of the Company began on 15 April 1923. This relay office was at Grimsby. Many similar offices were to follow throughout the Company's system.

Tillyer was justifiably congratulated by the Board. But there is no getting away from the fact that he was a lesser engineer than Higgins: it could scarcely have been otherwise. That the Company had survived the battle with the Press Association was due less to the architects of the Joint Service than to Higgins who had put the Company so far ahead of its rivals in means of communication. The gap was closing when the Company emerged from the difficulties of war. The first clear warning that it might be overtaken came in September 1919, when the Managing Director reported to the Board that the P.A. seemed 'about to embark on a scheme of news distribution by means of private wires and Creed apparatus'.

Wilfred King was well informed. To quote once more from George Scott's *Reporter Anonymous*:

> The system which the P.A. was to use after the war was the one invented and developed by F. G. Creed . . . In 1920 the P.A. installed the plant necessary to operate an experimental Creed transmission service over private wires to the

morning and evening newspapers in the West of England and South Wales. This called for the opening of the first provincial telegraph centre at Bristol.

The experiments were successful. The speed of transmitting P.A. news was revolutionised. A Swansea paper, the *Cambria Daily Leader*, told how the Creed had transmitted the text of a Lloyd George speech at Llandudno. The great man started speaking at a quarter to three in the afternoon. By the time the *Daily Leader* went to press at half past four it was carrying four and a half columns of the speech. As the P.A. Manager said, 'That would have been absolutely impossible over the public wires; we should not have dreamed of attempting such a thing'.

The Creed System was very briefly as follows. The news message was punched in Morse at the transmitting office. This punched tape was fed through a transmitter which produced an identical tape at the provincial centre. From there it was relayed on a circuit covering several provincial papers. The operators in these newspaper offices passed the tape through another machine which translated the Morse into clear. The speed of transmission was about one hundred words a minute.

It will be remembered that shortly before the war the Exchange Telegraph's Board had talked about the Creed System, then in an early state of its development. There had been correspondence with the Creed Bille Company, and one or two interviews. But nothing positive had been done. One feels that, had Higgins lived, he would either have produced something better or persuaded the Directors to hijack the Creed bandwagon. But on this occasion the Company was not the first to try the new system. They hesitated while a rival went ahead.

This must have put them at a disadvantage when a new Joint Service Agreement with the P.A. was being hammered out in the winter of 1921–2. In fact the use to be made of the P.A.'s Creed machines and private wires – the Association's System as it was called – constituted the main difference from the 1906 agreement, which had precluded instrumental delivery in the provinces. Clearly, of course, the workability of any such agreement must depend to a large extent on the clarity of the exposition. With that in mind, and having just been told that the saving to the Pool by the introduction of the Association's system shall be denoted as 'Z' in the document, one comes on the following sentence:

> The Association shall be entitled to receive from the Pool eighty per cent. of that portion of Z which bears to Z the same proportion as the number of Joint Centres for the time being connected with the Association's System bears to the total number of Joint Centres including London for the time being used for the purpose of the Joint Service.

The negotiators apparently understood this, for the new Service worked better than the old.

There were two other striking differences between the earlier agreement and the later one. The Company was given the right to open and manage its own provincial offices outside a radius of ten miles from any existing Joint Service office. Secondly, London Stock Exchange prices and Market Reports were also included in the Joint Service. This 1922 agreement was signed on 10 February.

At about that time the Board gave 50 guineas to a Testimonial Fund for Lord Burnham. What the particular occasion was is not clear, but the Company had good reason to be grateful to him. As the Hon. Henry Lawson, he had accepted the invidious task of acting as friendly adviser to the P.A. and the Company when they were growing tired of expensive litigation yet were still quarrelling.

One of the fascinating points about working at a history of this sort derives from stumbling upon incidents which wake associations and have been preserved in the archives because the Company was first in giving the news. There were two of these in 1919, a year which most people are too young to have experienced or too old to remember. The first was Harry Hawker's Atlantic flight. He as pilot and Commander Mackenzie Grieve, R.N. as navigator started from St John's, Newfoundland in a Sopwith-Rolls-Royce on the evening of 18 May. Their intended destination was the Galway coast 1,880 miles away. To lighten the aircraft Hawker dropped his

# THE TELEGRAPHIC NEWS

## OF

# THE EXCHANGE TELEGRAPH COMPANY.

ESTABLISHED 1870       PRICE 3ᴰ

7      MONDAY, NOVEMBER 13   1899.—(12.10)

## The War in South Africa.

The Exchange Telegraph Company states that the following telegram has been received at the War Office:—

" From the Officer Commanding in South Africa. Cape Town, 12th November, 8.45 p.m. Following message received from Nicholson, Buluwayo, 5th November: 'Following message received from Colonel Baden-Powell: " Mafeking, 25th October. All well here. Enemy still shelling us. We made a successful night attack on his advanced trenches last night, getting in with the bayonet. Our loss, six Protectorate Regiment killed, nine wounded, including Captain C. Fitz-Clarence, 3rd Royal Fusiliers (slightly), Lieut. Swindon. Names of killed: 4,323 Corporal Burt, 17th Lancers, 442 Trumpeter Iosia Lounday, 443 Charles Mayfield Middlebritch, 171 Thomas Fraser, 222 Alexander Henry Turner, 202 Robert Byris Macdonald. Enemy's loss unknown, but considerable. Enemy have now vacated Signal Hill, and laagered two miles north-east of town, and two miles south-east.' "

All applications for **Column Printing Telegraph Instruments,** Tape Machines and Telegraphic News Sheets should be addressed to the Secretary, 17 & 18, Cornhill, E.C.

Printed and Published by the Exchange Telegraph Co., (Limited), 64, Haymarket S.W., and at Brighton, Liverpool, Manchester, Edinburgh and Glasgow.

1   Historic issue of *The Telegraphic News*

J

THE FIRST CHAIRMEN
2 & 3 *(upper)* Lord Wm. Montagu Hay, 1872-1875;
Sir James Anderson, 1875-1882
4 & 5 *(lower)* Lord Sackville Cecil, 1889-1898;
Lt. Col. Francis F. Sheppee, 1898-1913

6  Captain William
   Henry Davies,
   Managing Director,
   1872-1898

7  *The Great Eastern,* built by Brunel

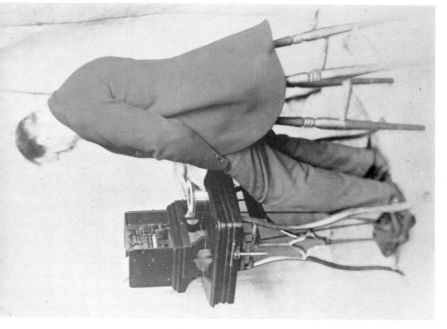

9 F. H. W. Higgins in early 1870's. Extel's first chief engineer with his column printer

8 Higgins' column printer news service instrument (weight-driven) 1887

Head Office for the United Kingdom:

COMPANY'S BUILDING

16, 17 & 18, CORNHILL, LONDON, E.C.

10   The Head Office of the Company was in Cornhill, from 1872 to 1919

## THE EXCHANGE TELEGRAPH COMPANY, LIMITED.

EDITORIAL PAY BOOK, *Week ending Saturday,* May the 7th 1898

| No. | Name. | Occupation. | Mon. | Tues. | Wed. | Thur. | Fri. | Sat. | Overtime Hours. | Overtime Amount. | Total Pay. | Signature. | Remarks. |
|---|---|---|---|---|---|---|---|---|---|---|---|---|---|
| 1 | Fish W.G. | Reporter | ✓ | ✓ | ✓ | ✓ | ✓ | ✓ | | | 2 15 | | |
| 2 | Serine J.E. | (Sub Editor) | ✓ | ✓ | ✓ | ✓ | ✓ | ✓ | | | 2 10 | | |
| 3 | Phillips C.M. | " | ✓ | ✓ | ✓ | ✓ | ✓ | ✓ | | | 1 15 | C.Phillips | |
| 4 | Whitelaw C. | " | ✓ | ✓ | ✓ | ✓ | ✓ | ✓ | | | 1 10 | Mabel Whitelaw | |
| 5 | Duckworth J.M. | " | ✓ | ✓ | ✓ | ✓ | ✓ | ✓ | | | 1 5 | J.M.Duckworth | |
| 6 | Brown C.J. | Gen Assistant | ✓ | ✓ | ✓ | ✓ | ✓ | ✓ | 26 | 5 5 | 17 6 | Chas.J.Brown | |
| 7 | Brown G. | " | ✓ | ✓ | ✓ | ✓ | ✓ | ✓ | | 1 - | 14 11 | G.Brown | |
| 8 | Black R. | Messenger | ✓ | ✓ | ✓ | ✓ | ✓ | ✓ | 6 | | 9 6 | R.Black | |
| 9 | | | | | | | | | | | | | |
| 10 | | | | | | | | | | 6 5 | 11 15 11 | | |
| 11 | | | | | | | | | | | | | |
| 12 | | | | | | | | | | | | | |
| 13 | | | | | | | | | | | | | |
| 14 | | | | | | | | | | | | | |
| 15 | | | | | | | | | | | | | |

Correct    John Born    Editor.

11   Weekly wage sheet from 1898

13  The last recruits, 1971

12  Extel messenger boy, c. 1894

*Alexandria Theatre*
*June the 28 1882.*

To

The Manager Liverpool Branch,
" The Exchange Telegraph Company, Limited,"
71, Lord Street, Liverpool.

Sir,

In reply to your Circular Letter of the *28th* of *June* I beg to state that *I* consent to *your making two porcelain fixings on top of the Alexandra Theatre for the purpose of Continuing a wire from Ganger Hall to London Road.*

on the understanding that the Directors of " The Exchange Telegraph Company," Limited, make good any damage which may be done to the property (referred to at foot) in the execution of the work, and to remove their Wires, Supports, &c.. at any time on the receipt of three months' notice to that effect.

Sir,
Your obedient Servant,
*Edward Saker*
*P. E. B.*

PROPERTY REFERRED TO ABOVE.

*Alexandra Theatre:*

14   The consent form for an old wayleave circular

15    1970, linesmen taking down the overhead wires on the
roof of the *News of the World* building

18 One of Extel's old tape machines with one of the latest, the DESCOM 200

16 Linesmen going underground to disconnect the underground cables

17 Curtain Road workshops near Liverpool Street Station, 1955

19  Sports day 1947, the wheelbarrow race

20  Sports day spectators, 1947

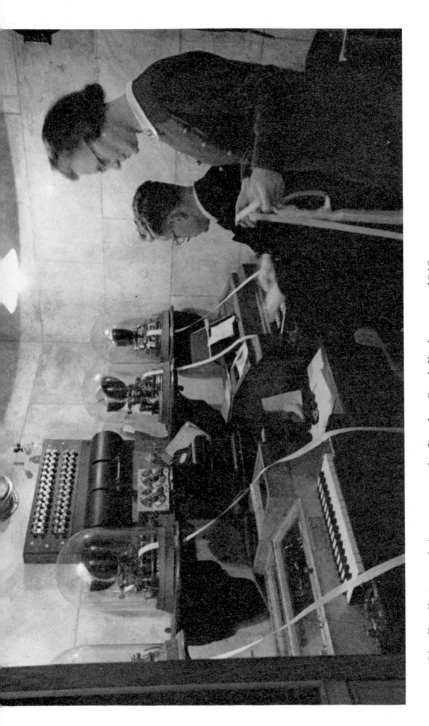

21  Extel's transmitting room at the London Stock Exchange, c. 1946

22   An Extel indicator board at the London Stock
Exchange, c. 1946

23   Extel reporters who work on the 'Floor' of
the London Stock Exchange, 1969

24   The Extel annunciator, now no longer in
service, in the House of Commons

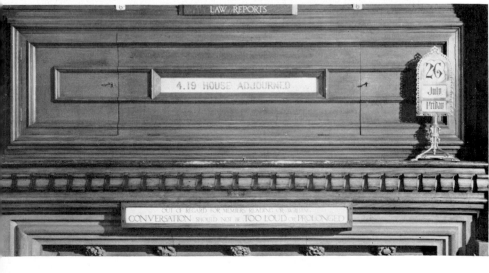

25    Switchboard operators of the Bartholomew House
       telephone exchange in 1956

26    And when this service was closed down 6 February 1970

27   The Manchester telephone room, 1953

Osborne
July 31. 1895

Dear Sir,

Now that the General Election is practically over I have much pleasure in writing to thank the Exchange Telegraph Company, on behalf of the Queen, for the arrangements so kindly made by which Her Majesty has had the earliest possible reports of the result of every Election together with

28 Part of a letter from Queen Victoria's Private Secretary

29 Handling General Election Results, October 1959

30 Mr Edward Heath studies the latest news from Extel at the British Industries Fair in Barcelona, April 1964

31   King George VI in North Africa with Extel war correspondent, Edward Gilling

32    Queen Elizabeth the Queen Mother visits the sports room at Extel House

# The TELEGRAPHIC NEWS

15   OCTOBER  3,  1903 (6.0)   Price 3d

Printed and Published at 12, Middle Street, Brighton.

# FOOTBALL.

## FIRST  LEAGUE.

| | | | | |
|---|---|---|---|---|
| Sheffield Wed. | . | 2 | Notts Forest | . | 1 |
| Sheffield United | . | 1 | Manchester City | . | 0 |
| Wolverhampton | . | 2 | Notts County | . | 0 |
| Liverpool | . | 2 | Small Heath | . | 1 |
| Sunderland | . | 2 | Blackburn Rovers | . | 0 |
| Newcastle United | . | 3 | Derby County | . | 1 |
| Everton | . | 2 | Middlesbrough | . | 0 |
| Stoke | . | 2 | Aston Villa | . | 0 |
| West Bromwich | . | 3 | Bury | . | 2 |

## SECOND  LEAGUE.

| | | | |
|---|---|---|---|
| Woolwich Arsenal | 4 | Manchester Un't'd | 0 |
| Barnsley | 2 | Bristol City | 0 |
| Lincoln City | 1 | Burton United | 0 |
| Gainsborough T. | 4 | Stockport County | 1 |
| Chesterfield | 2 | Blackpool | 1 |
| Bolton Wanderers | 3 | Leicester Fosse | 1 |
| Preston North E'd | 3 | Burslem Port Vale | 1 |
| Burnley | 2 | Grimsby Town | 0 |

## FIRST  SOUTHERN  LEAGUE.

| | | | |
|---|---|---|---|
| Millwall | 7 | Kettering | 1 |
| Portsmouth | 1 | Swindon | 0 |
| Reading | 2 | Tottenham Hotsp'r | 2 |
| Bristol Rovers | 5 | Brentford | 1 |
| Plymouth Argyle | 0 | Luton | 0 |
| Northampton | 4 | Brighton & Hove | 0 |
| Queen's Park R. | 3 | New Brompton | 0 |
| Fulham | 2 | Southampton | 2 |

34   Charles Wills,
General Manager,
1928-1949

35   An Extel commentator at a 'heath' meeting at Ascot

36   The trophy and wine cup mementos for the
£10,000 Extel Stakes at the Goodwood July
meeting

37   1968—demonstration of tic-tac numbers
1 to 6 (reading left to right)

# THE EXCHANGE TELEGRAPH COMPANY, LIMITED.

## Incorporated 1872,

*Under the Companies Acts, and a special license from Her Majesty's Postmaster-General.*

## CAPITAL, £225,000.

### HEAD OFFICES—17 AND 18, CORNHILL, LONDON, E.C.

THIS Company is now establishing Branch Offices or Call Stations throughout the Metropolis and Suburbs for the due administration of its

## TELEGRAPHIC CALL SYSTEM,

By the aid of which, subscribers may be enabled at any hour of the

## DAY OR NIGHT

### TO CALL A

## MESSENGER, CAB, or POLICEMAN,

### AND GIVE THE

## ALARM OF FIRE,

While many other "Calls," indicating the wants of a private house, chambers, office, or place of business of any kind, may be arranged for, all such being made in the same uniform manner by the pressure of a button on a small automatic instrument, placed as most convenient, and telegraphically connected with the nearest "Call Station" of the Company, which in no case will be distant from the subscriber, more than a quarter of a mile, or three minutes' time. These instruments occupy but a few inches of space, are not liable to get out of order, require no local batteries or winding up, and no knowledge whatever of telegraphy to work them.

## CALL STATIONS

Will be established wherever a demand for them may arise ; they will be provided with a permanent staff of Operators and Messengers, whose duty it will be to receive and attend to "Calls." A policeman will be found there, and an expert with a hand-pump or extincteur ready to act on the first alarm of fire, and each Station will be in telegraphic communication with the nearest

## POLICE AND FIRE BRIGADE STATIONS,

Thus enhancing the public value of the system by the increased security which will be rendered to life and property.

The system has been in operation in the United States for some years past, having been initiated and established by the American District Telegraph Company to meet a great public want, and where its practical advantages have become so fully recognised and appreciated that it is being rapidly adopted throughout the States ; in New York, where it originated, and some thousands of instruments are at work, it is considered indispensable as

## THE HOME TELEGRAPH.

A PROTECTION TO LIFE AND PROPERTY—A CONVENIENCE IN DOMESTIC LIFE, AND AN ADJUNCT TO BUSINESS.

## DAY SERVICES,

Chiefly for the purpose of utilising the Messenger Service, will be established in business localities

### TERMS FOR A SINGLE INSTRUMENT :

Day and Night Service ... ... 5 Guineas per Annum.
Day Service ... ... ... ... 4 Guineas per Annum.
### Extra Instruments at a Reduction.

Instruments placed and connected with the nearest Call Station of the Company, and kept in working order free of all charge to subscribers.
Messengers, when employed, Sixpence to Eightpence per hour.
Automatic, Fire, and Burglar Alarms fitted in conjunction with the Call System.

*For further particulars, apply to the SECRETARY, at the above Address.*

38    An advertisement in 1876 for a former Extel service: the telegraphic call system

39 Modern Extel showing some of its wares at the
Computer '70' Exhibition at Olympia

40 Extel stand at the 1965 Bookmakers and Betting
Shops Exhibition in London

*EXTEL BRITISH COMPANY SERVICE*

## EM - EZ 25    EXCHANGE TELEGRAPH COMPANY (HOLDINGS) LTD. (THE)    EXC

**NEWS CARD**    PLEASE WITHDRAW PREVIOUS NEWS CARD.    UP-DATED TO 27-7-71.

ISSUED EQUITY CAPITAL: £1,773,285.

GROSS YIELD INDICATOR based on Dividend 22%, Earnings 39·0% (9·75p per share), Capital £1,773,285.

| PRICE | p | 120 | 130 | 140 | 150 | 160 | 170 | 180 |
|---|---|---|---|---|---|---|---|---|
| DIVIDEND YIELD | % | 4.58 | 4.23 | 3.93 | 3.67 | 3.44 | 3.24 | 3.06 |
| EARNINGS YIELD | % | 8.13 | 7.50 | 6.96 | 6.50 | 6.09 | 5.74 | 5.42 |
| P/E RATIO | | 12.31 | 13.33 | 14.36 | 15.38 | 16.41 | 17.44 | 18.46 |

SHARE PRICES. 25p ORD. (LONDON): *1971, Highest 125p, Lowest 100⅝p. *To June 17.

ORD. DIVIDEND PAYMENT DETAILS. Year end March 31. Last accounts published 30-6-71.

| | | %<br>Per Share | Announced | Paid | Holders | Ex Date |
|---|---|---|---|---|---|---|
| 1970 | Int 8 | 4.8d. | 20-11-69 | 19-12-69 | 2-12-69 | 24-11-69 |
| | Fin 12 | 7.2d. | 28-5-70 | 23-7-70 | 22-6-70 | 1-6-70 |
| 1971 | Int 8 | 4.8d. | 26-11-70 | 21-12-70 | 2-12-70 | 30-11-70 |
| | Fin 14 | 3½p | 27-5-71 | 22-7-71 | 25-6-71 | 14-6-71 |

GROUP HALF YEARLY FIGURES TO SEPT. 30 (Unaudited).

| | GROUP PROFIT | DEPRECIATION | PROF.BEF.TAX | TAX | PROF.AFT.TAX |
|---|---|---|---|---|---|
| 1968 | £724,000 | £116,000 | £608,000 | £260,000 | £348,000 |
| 1969 | £644,000 | £129,000 | £515,000 | £212,000 | £303,000 |
| 1970 | £715,000 | £145,000 | £570,000 | £231,000 | £339,000 |

26-11-70.

INTERIM REPORT. Figures for half year to 30-9-70, included in table.

27-7-71.

AMERICAN DEAL. It is announced that Co. and Inforex, Inc., an American based manufacturer of computer terminal and data processing equipment, have reached agreement in principle under which Co. will be appointed the exclusive distributor for Inforex Key Entry Equipment in the U.K. and Ireland and will provide sales and service support. This operation will be handled by Co.'s Data Systems and Engineering Divisions.    Agreement is subject to Co. acquiring the Inforex related interests of Computer and Systems Engineering Ltd., the previous distributor.

---

41    Sir Wilfred Creyke King, Chairman, 1913-1943

42    Brig. General S. M. Anderson, D.S.O., Chairman, 1943-1954

43    'Stats' have been producing their cards for just over 50 years. Here is part of the card for the Extel parent company

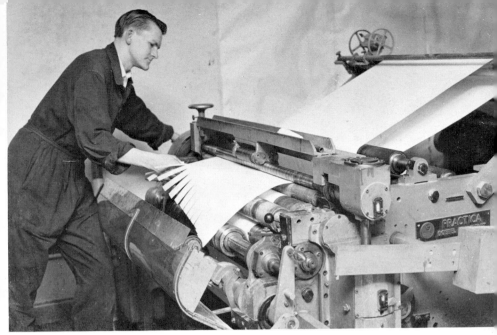

44   Thames Paper Supplies, a member of the Extel Group, produces paper tapes and rolls for data processing and telecommunications. A slitting machine operator at work

45   The Extel electronic annunciator system at St. Pancras, 1969

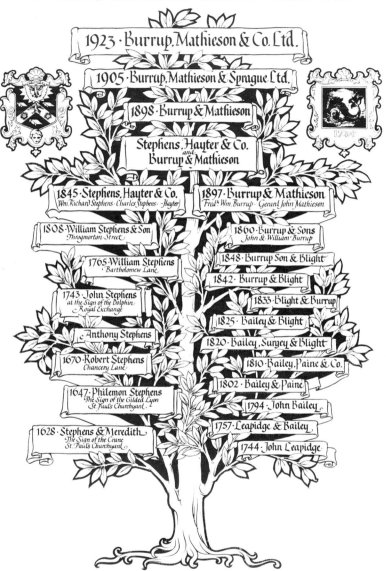

Three Hundred Years – and on

1923 · Burrup, Mathieson & Co. Ltd.

1905 · Burrup, Mathieson & Sprague Ltd.

1898 · Burrup & Mathieson

Stephens, Hayter & Co.
and
Burrup & Mathieson

1845 · Stephens, Hayter & Co.
Wm. Richard Stephens · Charles Stephens · Hayter

1808 · William Stephens & Son
Throgmorton Street

1765 · William Stephens
Bartholomew Lane

1743 · John Stephens
at the Sign of the Dolphin
Royal Exchange

Anthony Stephens

1670 · Robert Stephens
Chancery Lane

1647 · Philemon Stephens
The Sign of the Gilded Lyon
St. Pauls Churchyard

1628 · Stephens & Meredith
The Sign of the Crane
St. Pauls Churchyard

1897 · Burrup & Mathieson
Fredk. Wm. Burrup · Gerant John Mathieson

1860 · Burrup & Sons
John & William Burrup

1848 · Burrup Son & Blight

1842 · Burrup & Blight

1833 · Blight & Burrup

1825 · Bailey & Blight

1820 · Bailey, Surgey & Blight

1810 · Bailey, Paine & Co.

1802 · Bailey & Paine

1794 · John Bailey

1757 · Leapidge & Bailey

1744 · John Leapidge

46    Burrup's Family Tree

47 *(top left)* William Charles Stevens, Chairman, 1954-1961

48 *(top right)* Thomas F. Watson, Chairman, 1961-1968

49 The Board of Directors of The Exchange Telegraph Company (Holdings) Limited, July 1971. *(l. to r.)* Ernest W. H. Bond (Secretary), Leslie A. Cross, John P. R. Glyn, John L. Harvey, M.B.E., D.L. (Deputy Chairman), Glanvill Benn (Chairman), Alan B. Brooker (Managing Director), Samuel P. S. Bartlett, Derek G. Lee

50   Extel House by night

undercarriage and wheels into the sea immediately after take-off. The plane did not reach Ireland, and there was no news for a week. But a cable ship sighted its red light on the water on 19 May and a Danish vessel rescued the two men after they had been ninety minutes in the water, 1,100 miles from Newfoundland and far out of their course. The Company picked up the news from this vessel on 25 May.

The other incident was the heavyweight Championship of Europe fight between Joe Beckett and Georges Carpentier on 4 December. The Exchange Telegraph man, André Glarner, must have been as quick on his feet as the French boxer, for the fight lasted only a few seconds, Beckett being knocked out by the second blow – a right hook to the chin following a straight left to the body. It looks as if Reuters were piqued by this Exchange Telegraph success, for in *Reuter's Century* Graham Storey says that when Dempsey knocked out Carpentier in Jersey City in 1921 their reporter got the news across the Atlantic to London in two seconds.

In this context, both of subject and of time, the name of Charles Wills may be appropriately introduced, for Wills had an ardent interest in sport, and in July 1919 he was made Assistant to the Managing Director. Here was Wilfred King shedding some of his load and sharing responsibilities! It was a great event. But Wills had been well tried. He was thirty-nine, had been with the Company since he joined it as a junior clerk in 1897, had been in both the sport and financial departments and had worked at the drafting of the 1906 Joint Service Agreement.

At the end of 1919 the Board decided 'to invite the staff to dinner to meet those members who have served in H.M. Forces'. There is only this bare statement on the record, and it makes one wonder how this well-meant party went off. The question, spoken or unspoken, 'What did you do in the Great War, colleague?' must have been in many minds. On the one hand were those who had borne under increasing difficulties the unglamorous heat of the office day for over four years and might be wondering how their hard-won positions would be affected by this mass increase to the staff. On the other were those who had borne the dangerous heat of war and might be

L

wondering exactly what it would mean to get their old jobs back. Nearly all the guests must have known each other already. Did this invitation to the two categories to 'meet' each other accentuate the difference unnecessarily? There must in fact have been harmony, for otherwise there would have been some note or hint of it: the Board had a passion for analysing anything that went wrong – which the Staff must often have found tiresome. But not a word. One supposes that the family feeling which senior members of the present staff, active and retired, still talk about with warmth was strong enough to cement the reunion.

On the business side there was enough expansion and innovation to keep a larger staff occupied and interested. To give only a few varied instances:- the statistical and advertising agency department continued to go from strength to strength. An office was opened in Copenhagen. The Manchester office required more accommodation, as also did head office, for they began to look with envious eyes at No. 62 Cannon Street, next door. The L.C.C. ordered twelve annunciators for the new County Hall. A column printing instrument was installed at 10 Downing Street 'free of charge': one does not know to what extent this was a peace offering to Lloyd George after his libel action. An extra room was acquired in Bartholomew House and another clerk added to the Company's Stock Exchange team. Lathes and drills driven by electric power were installed at the Creechurch Lane workshop.

The Company was determined not to be left behind if wireless telegraphy and telephony developed into a useful means of disseminating news. But there is paradox in the history of radio: this swift thing was so slow to be exploited in the popular field. In December 1901 Marconi sent his first signal across the Atlantic. In 1908 the Board of the Exchange Telegraph studied with interest a report by the Milan correspondent of *The Times* about official trials of the De Forest wireless apparatus which had been carried out between two ships of the Royal Navy in the Bay of Genoa. Signals in words, not Morse, had been successfully passed between the two warships when twenty miles apart. And there we were in the 1920s with broadcasting a new thing and sadly unorganized.

There was, of course, the 1914–18 war which hid developments under the thick cloak of security. But thereafter everyone
was slow, and Britain slowest, in making efficient public use of
the 'new' invention. In one way this was to Britain's advantage.
She learned by the mistakes of others. In the United States,
where broadcasting for entertainment began a couple of years
before it did in Europe, there were between five hundred and
six hundred transmitting stations, eleven in New York alone,
and in the early days they all used the same wave-length. This
was soom amended, but there was only one government
service. Most of the other transmitters belonged to electrical
companies, department stores, newspapers and radio societies.
Apart from the last named, which received a subscription
from their members, all this listening estimated at three
million sets was free – there was no licence to buy. The transmitting sponsors advertised their own products or those of
others at a fee, thus covering expenses.

For some whimsical reason the Postmaster-General was in
Britain the person endowed with the responsibility for
organizing radio. However efficient the G.P.O. is now, it did
not do well with earlier means of quick communication, such
as the telephone. (The Press Association, as has been seen, stole
a march on the Company largely by acquiring a com-
prehensive service of private lines.) Very early in the 1920s
some organizations were ready to spend big money on broadcasting. But without the sanction of the Post Office no one
could legally listen-in ('set up an experimental listening
station' was the official phrase), let alone transmit.

The first White Paper on the subject is charmingly naïve
in its use of what seem to us now very obvious statements:

> The outstanding feature of radiotelephony is that it
> enables a single voice to reach innumerable ears . . . It may
> be that broadcasting holds social and political possibilities
> as great as any technical attainment of our generation.

Beginning in 1920 and continuing throughout the decade
there are numerous references to wireless and to broadcasting
in the Company's minutes. Most of them are in the usual
unenlightening form: a letter was read to the Board or the

Managing Director reported on a meeting. But the keen interest, even anxiety, of the Board is evident, and a few instances in the early years give a clue as to the way their minds were working.

In the early summer of 1920 they were invited to consider various press articles about broadcasting and 'listening-in apparatus'. These are not worth quoting now. In July the Managing Director reported on an interview with Mr Binyon of the Radio Communication Co. Ltd., 'and on arrangements made for the Listening-in trial'. Presumably this referred to 'an experiment in wireless telephony' which was made at Cannon Street on 26 August. Tantalizingly, one can find out nothing more. But one's mind's eye seems to see the Chairman and Managing Director, a true blue Conservative in all things, who looked like King Edward VII, wore a top hat and still wrote a double-S like a copybook f, a non-scientist and then aged fifty-nine, trying to assess the potentiality of a talking box.

Later he went still further. He bought 'a listening-in apparatus' (no doubt a crystal and cat's whisker set) for £39 3s 0d, charged it to the Company and had it installed in his house. Then he told the Board what he had done, adding that the set was and would remain the property of the Company and that he would experiment with it. The Board approved. So we have another picture, of W. K. turning the knobs and listening to the crackles and squeaks in his bachelor home – 'The Chalet', Roehampton.

There are a number of references to letters from and conversations with members of the Marconi Company. Marconi had a government monopoly for wireless throughout the war, and they maintained it until the Post Office took over. One wonders if the Board were exploring the possibilities of a deal. The situation is fascinating: here is the Exchange Telegraph, after fifty years of life, facing a situation closely analogous to that in which it began. Then the trans-ocean telegraph companies had but lately started to operate. Now there was another means of spanning the seas.

There seemed to be three ways of making money out of broadcasting – apart from transmitting oneself: by supplying commercial information such as Stock Exchange prices, by supplying general news, and by advertising. (It may appear that the differentiation between the first two is too subtle. But it still remains. Sport and all else except financial information is included in the general news. But City prices are held back until a late hour when children of all ages are presumed to have gone to bed or the night clubs. There are recurrent objections to sex and violence being made available to the young by radio or television, but the Market Stock Exchange prices from the Exchange Telegraph and Reuters are kept well out of their reach.) The Government of the day set its face sternly against advertising by wireless. So that was that.

But the supplying of news, general and financial, remained a possibility. And looming larger was the fear that copyright news would be pirated. Several instances of news-stealing have been given. They were scarcely a tenth of the whole. And we were dealing with telegraphic news where stealing depended upon considerable skill and plotting. If news, the daily bread of the agencies, was flying about all over the ether, how could one possibly control poachers?

The Exchange Telegraph was not alone in its anxiety about the new medium, nor was its chief worry the only worry. The newspapers were concerned that if news was broadcast virtually free they would lose their purpose, and their money. The point of view of the agencies was more complicated. If the newspapers, their main customers, lost money they would lose in proportion. But on the other hand if the broadcasting organization was ready to buy news, and only the agencies supplied it, here was a new market.

The Marconi Company never entered the competition. They stated very clearly that they would stick to carrying messages and avoid all forms of news reporting. On 15 November 1922 a daily broadcasting service was started from London, with a few relaying stations which were later extended. This was done by the British Broadcasting Company Ltd. which, with Government approval, had been formed among manufacturers of wireless apparatus. They paid a royalty of £50 to the P.M.G. for each transmitting station, and anyone could listen-in if he paid 10s for a licence from the Post Office and bought a British-made set; or alternatively if they made a set themselves, for that (it was thought) would prove they were sufficiently skilful to be useful as experimenters.

The most fascinating point about all new legislation is the way its carefully phrased terms are immediately circumvented. In this case there appeared upon the market foreign-made parts which any fool could assemble into a receiving set. This is nothing to do with the Exchange Telegraph, but it helps to set the complex scene in which the Company was then involved.

The agencies and the newspapers got together and laid down, not the law, but their influential conditions. Broadcasting must keep to its legitimate field – comment and entertainment. Nothing which might be defined as news should be broadcast before seven p.m. by which time the evening papers had sold their wares. These negative and impracticable restrictions must be appreciated in the atmosphere of thought at the time. It was logical enough. If people could have news for nothing, or next to nothing, why should they pay for it at the cost of a daily newspaper? News was a commodity, to be sold, not stolen or picked up for nothing in the street.

This all falls five years beyond the period of this chapter, but precisely for that reason it accentuates how obstinate and long-enduring was the objection to news being scattered without control. By arrangement with the B.B.C. the news had been prefaced since December 1922 with the sentence, 'Here is the news, copyright by Reuter, Press Association, Exchange Telegraph and Central News'.

In 1927 the Company sought counsel's opinion as to whether their copyright had been abused by a shopkeeper who had a loudspeaker blaring outside his premises, thereby allowing the public to listen to the news for nothing. The K.C., for a fee of one hundred guineas or so, gave it as his opinion that:

> In order to get the loudspeaker to operate the licencee must do something himself, viz – he must acquire and instal his receiving instrument, and set and tune it to the wave-length of the distributing office, and he must also

connect up the loudspeaker with the receiving instrument. Without these operations on his part the loudspeaker would not function, but when it does function and it can only do so in result in part of the licencee's operations, it reproduces the words of the broadcast messages as they are transmitted by the Broadcasting Corporation [the B.B.C. was by then a Corporation] and received by means of the receiving instrument. No copy of the message has to be made nor is it made for the purpose of the licencee. The copyright messages are therefore not infringed by copying.

Who nowadays would think of suing the shopkeeper except possibly for the nuisance of noise? The idea that broadcast news and printed news must compete together, not complement each other, died hard. In both Canada and the United States the press and the radio companies clashed head on. That nothing so violent occurred in Britain was due largely to the foresight and tact of John Reith, the father of the B.B.C. He foresaw that nothing could hold back the development of broadcasting in the long run. Therefore he was conciliatory in his dealings with the press and the agencies. They did not want the B.B.C. to carry advertising: nor did he, so he agreed with them. They demanded that there should be no news broadcasting before seven p.m.: he initially agreed to this, and only when public opinion was on his side did he advance the hour. The papers insisted that the B.B.C. must pay for the publication of its programmes: Reith refused to do this, but the papers were soon forced by their readers' demands to publish the programmes as editorial matter.

The agencies, of course, had little or no reason to complain. They were paid a lump sum which varied according to the number of receiving-set licences. But if the agencies did not quarrel with the B.B.C. they squabbled among themselves about the division of the royalty.

In 1922 this extra source of revenue had not yet begun to flow. But the Company made a profit of £17,000 and paid a ten per cent dividend. It bought the freehold of 62 Cannon Street (next door to 64) for £13,000, which was considered favourable terms. At the General Meeting £2,500 was voted as a bonus for division among the staff according to length of service. (One man, Thomas Stokes, had been with the Company for forty-nine years.) The Company and the P.A. achieved their most popular joint service by renting a sports ground. All in all the jubilee year was the best to date.

CHAPTER 9

# A Changing World

AT OVER fifty the Exchange Telegraph was getting to the age when the passing of old friends became a sadly frequent occurrence. Thomas Stokes who had joined the Company in 1873 died in harness on 20 May 1924. Wilfred King whose length of service was not much less attended the funeral, and the Board voted £50 to the widow.

During the next few years there were several other casualties among the long-service men – A. Taylor, foreman linesman, who had joined in 1880, George Ward who had been a mechanic for the same period, and A. W. Collis, wayleave clerk, who had been in the Company for thirty-seven years. The minutes record only that the Managing Director 'interviewed' Mrs Collis, but elsewhere there is a note to the effect that she wrote a letter of thanks for the gift of £100. In 1929 the head printer, A. Shrimpton, died at work, aged seventy-six, after forty-three years' service.

It is sad that one should have heard nothing of such men until they died. But it is apt to be like that even with men who in their prime were well-known figures, but whose record one only learns from the obituary. And even then the character, as opposed to achievements, is rarely touched on. To avoid that being the case with Wilfred King, we will attempt a sketch of him as he was when still going strong. The material exists, for there are members of the staff – active as well as retired – who served under him at this period. The raconteurs vary from men who held very junior positions at the time to one or two who were sufficiently senior to become Chairman themselves later on. So one gets a varied point of view.

In the early days – at least until the 1922 Jubilee and

probably later – the Managing Director interviewed everyone on the day he joined the Company. The scene was the Board-room which the Managing Director used as an office – a large room with a desk, a table, a polished floor with a few mats on it and a screen which was there for no apparent reason except that screens had been in fashion at some time during the last fifty years. At the desk sat a bearded Victorian – or Edwardian – gentleman who for some time went on with his work apparently unconscious that anyone was standing nervously in front of him. At last the massive head was raised. Questions were asked, a task appointed.

The impact of that first interview lasted through perhaps forty or more years of work for the Company, and into retire-ment. There were bound to be later interviews as well, for King liked to know everything, and at first hand. If the subject to be enquired into was known or guessed at, the person concerned naturally prepared his answer. But the Managing Director had a way of asking some totally unexpected question as well. He was an exacting master. He hammered into every-one that seconds mattered. If anything was late he held an inquest.

In his own movements he was slow. He never walked quickly, never appeared to be in a hurry. But, as an ex-linesman puts it, no one could move faster if someone was in trouble. A doctor, a nurse – whatever was necessary was provided at once. And although King never passed over a mistake he always said 'thank you' for good service rendered.

Particularly was he a martinet in matters of dress. A messen-ger boy might find himself stopped in the street and severely reprimanded for having one button of his jacket undone. A sub-editor who prided himself on his elegance once wore in the office the new fashion, a soft collar. He was told never to appear again 'with that rag round your neck'. There is another side to that coin. One poor man – poor in both senses – turned out to play cricket in dark trousers and with a borrowed bat. He received an anonymous present of a complete outfit.

These are little things. In executive duties King was quite as punctilious and precise. He did not like anyone to argue with him, even a Director. With regard to the Company's varied interests, the Managing Director attended to everything as a duty. His numerous and extensive tours show how closely he kept in touch with provincial offices and subscribers of every category. But his heart was not in sport – racing, cricket, football. Probably he was happiest when he went off in his top hat to the Stock Exchange, or to interview some senior government official, or travelled to Geneva to attend a press conference of the League of Nations. Perhaps also when he was driving a bargain with his opposite number in the Press Association!

No example has been given of his sense of humour. There are not many, and none suggests subtlety. A lady employee was on the mat for being late on many occasions. This was something for which she might well be sacked, and she entered the Boardroom with the suddenness of fear. She stepped on a mat which slipped on the polished floor, and she sat down very suddenly in front of the desk. The General Manager went behind the screen where he apparently suffered from a severe fit of coughing. The lady was not even reprimanded.

There was, of course, 'No Smoking' in the office. Wilfred, as he was called – well behind his back – was sufficiently omnipresent in head office for this to be strictly observed. But the editorial, which was tucked away in Panton Street, and the writers felt secluded enough to allow themselves a little latitude. It is said that when the Managing Director unexpectedly appeared you could see, if you looked carefully, little wisps of smoke rising from the drawers of office desks. It is also said that he once sacked a visitor for smoking. But this sounds apocryphal, for he knew everyone.

Practically nothing is known about the home life of this bachelor, non-smoker and non-drinker. Douglas Anderson remembers imposing tea parties at the Ritz for the younger members of the founder's family. Then Abdulla cigarettes were handed round. It was a great occasion.

And every summer, except of course during the war, and later in life when he went to his house in Deal, King spent his holiday in Switzerland. Did he there commune with the mountains? Had he the philosophy of a mountaineer? No one can answer. The big man in a little office world was the only person in his private world. One cannot help wondering if he felt lonely.

There were changes on the Board. In 1923 Kenneth Belben Anderson, one of the two sons of the Kenneth Anderson who died in 1914, was elected to the Board. Thus the family was once more represented after a gap of nine years. K. B. Anderson was still a young man. In the files there is a letter from him dated 9 March 1918 when he was a Royal Artillery lieutenant. He wrote to King:

> As you know I am still a fairly large shareholder of the Exchange Tel. Coy. and to tell the truth I am not a little worried about it. I have been advised to get out of it 'if possible' but from old associations' sake I shall be sorry to do that . . .

Far from getting out, he became a Director four and a half years later. There seemed every prospect of him remaining so throughout an active life, but in June 1925 he was granted six months' leave of absence due to ill health, and on 12 November he died.

This time there was virtually no break in the representation by the founder's family. Brigadier-General Stuart Milligan Anderson, D.S.O., was voted a Director within a month. Yet it does not seem that he was merely appointed in his brother's place, although for the first three years he was a non-executive Director. There exist some manuscript notes by Wilfred King which suggest (they do nothing so injudicious as to state) that by 1928 he was looking for a successor as Chairman. It is clear that the soldier needed some persuasion to give his life to business, and there is the implication that a worthwhile prize was offered. Since Wilfred King was sixty-seven it might have been supposed that he would retire within a few years.

It is a sound literary principle that one should not give away in advance the end of any story, whether fiction or fact. But the reader may picture a senior military officer brought into something like the family firm on the tacit understanding that he would soon command it. There is independent evidence that this was so. A person who actually saw the scene resulting says that General Anderson shared an office with King, but King always went out to the room of his secretary to dictate anything confidential.

The minutes show that General Anderson was often sent to interview people, for example the representative of Creed. His report was read to the Board. But no resolution was taken until the Managing Director had personally looked into the matter.

This apparent lack of confidence should not be taken as a criticism of Wilfred King. There was no fault of selfishness, pettiness, suspicion. But it was the mettle of the man to hold all the reins. It was almost impossible for him to hand over even one pair while he still felt himself capable of guiding the business as no one else possibly could – which was, perhaps, true. After all he was not only the Chairman and Managing Director, he had far more experience than anybody else. The Manager, Charles Wills, who had been with the Company for nearly thirty years, was given scarcely more freedom than Anderson.

Horace L. Hotchkiss had been a Director almost from the Company's inception. His home, 100 Broadway, New York, was so far from London that he was excused from attending Board Meetings. But he gave his attention to business on the American side and looked after any colleague who crossed the Atlantic. On 11 March 1928 Horace Leslie Hotchkiss resigned from the Board. At the meeting when this was announced his son, Horace Leslie Hotchkiss Junior, was elected in his place, he like his father remaining in America and not attending Board Meetings. The researcher into the Company's history scarcely notices any difference.

Another resolution of this same Board Meeting reads as follows:

That in view of the reorganization of the Staff the Directors taking into consideration the age of the Secretary, Mr George Ferris Hamilton, and the necessity of providing for the future, deem it well to place him forthwith on the retired list and that in view of his long service with the Company he be granted an allowance at the rate of £700 per annum for the next three years as from 1 July 1928 and thereafter at the rate of £500 per annum.

Hamilton's contract of February 1906 provided for six months' notice. At the next meeting he was given twelve months' full salary in lieu of notice.

Hamilton's successor, William Broad, resigned within a year and he was succeeded as Secretary by William Charles Stevens, the Assistant Secretary.

There was even more activity than usual in the engineer's department. Both King and Tillyer visited the United States in search of printing instruments. Kleinschmidt instruments were first ordered from Chicago. The Board showed as much excitement about them as a father might over the birth of a child. The minutes recorded which liner was bringing them over, their arrival at Southampton, then at Newcomen Street, S.E.1 – (the enlarged engineering quarters which had been taken in place of those in Creechurch Lane), their assembly, trial, instalment. Kleinschmidt instruments recorded the opening of Parliament in January 1925. Only Tillyer, the engineer, was less than enthusiastic about them. He reported:

Our Daily Service has been equipped with twenty-three Kleinschmidt instruments in order to increase the speed of transmission. These instruments are giving good service but are very complicated and expensive to maintain, as would be expected by reason of speed and the fact that they are not of British make.

In the tradition of Higgins, Tillyer seems to have been right, for the Kleinschmidts were gradually replaced first by Morkrum-Kleinschmidts and then by Creeds.

Great strides were made in increasing the underground linkage in place of overhead wires. The major scheme was to link Panton Street with the underground wire in Pope's Head Alley behind Cornhill, and on to Cannon Street and the Stock Exchange. It would then be possible to transmit directly from the Stock Exchange – 'should it be thought advisable, a matter well worth consideration,' wrote the engineer to the non-technical Board. It was hoped to finish this work before the end of 1925, but it was held up by rebuilding on the Cornhill site of the handsome Lloyds Bank premises we see today.

Underground lines on this and other circuits were crowded

in with gas pipes, Post Office wires, sewage pipes. There did not exist the commodious underworld London of today where one can walk upright, say, along the line of Charing Cross Road, passing side tunnels signposted with the names of the streets above.

The obvious advantage of getting the lines underground was that they were safe from wind and snow. But they still had enemies – corrosion and rats. Corrosion could only be countered by constant inspection. The inquisitive nature and powerful teeth of rats could not be countered at all, having been ordained by a higher authority. Now and then a rat explored a tube. Rats cannot walk backwards because their hind legs are longer than the front legs. So they pressed on until they stuck. Then, growing hungry, they gnawed through the gutta percha which still insulated many of the wires. They bit further and electrocuted themselves. Served them right, perhaps, but they broke the circuit.

Tillyer in one report gives an idea of the difficulty of laying these lines under the busy City streets. Since the Company's engineers had no statutory powers they could only do this work between Saturday afternoon and early Monday morning. Tillyer never missed an opportunity to praise his staff. There is no record of complaint about those lost week-ends. And in those days overtime was paid no more than normal working time. This was fifty-three and a half hours a week, reduced in 1929 to forty-seven hours.

Business was brisk. In November 1924 over 12,000 phone calls were made by members of the Stock Exchange through Bartholomew House in one day. Five years later the total exceeded 18,000 with 1,500 'after hours' calls to be added. The Joint Service was working better than it ever had before. The Company's Board endorsed the suggestion of the Joint Committee that the agreement should be extended by twenty years. More joint service stations and provincial offices were set up. That of course meant more work for the engineers. And on top of all this they suffered fire and high water.

There were four subway fires caused – according to Tillyer – 'by faulty cables belonging to the L.C.C.' But the Company was instructed to remove all wires insulated with gutta percha (in the City area) and to lay the new lines – about fifty miles of lead-insulated cable – in iron troughing.

Then, on 27 December 1927, two days after a snowstorm which put fifty-one circuits out of action, the river overflowed the Embankment and flooded the L.C.C. subway. This happened at two o'clock in the morning, and some people who lived in basements were drowned in their beds. Linesman Percy Davies was nearly added to the list of fatal casualties for he happens to have no nose for smells, and when he went down to look for faults he was overcome by gas from fractured pipes. There were faults in plenty, the test boxes having been put out of action. By diverting some of the circuits which were not in use at the time, Tillyer's men got everything working again within a fortnight. But they had to re-lay twenty-three miles of wire.

A minute of 9 January 1930 states that 'Mr E. G. Tillyer had been advised that his employment with the Company would terminate on 31 January'. He was given £2,000 'for being deprived of his office of engineer to the Company'. E. Steadman was appointed in his place with F. C. Wolstenholme as assistant engineer.

This is the second instance given of something which evidently went wrong but for which there is no explanation. It might be argued that such things should be omitted from a history: everything mentioned ought to be cut and dried. That is not so. It must have been as evident to the reader as it is to the writer that this story has so far developed more as the annals of a family than a Company history. Without exception everyone alive who was on the staff in the late twenties and early thirties has remarked on the close personal relationship which existed. 'We were still a small company; we were a sort of family' is a typical response to questions about discipline, applications for a rise of salary or wages, ideas on development – almost anything. Of course there were happenings which seem unsatisfactory to people far removed in time. Every family has its secrets. It talks willingly about its successful sons but skates over behaviour which has pleased it less. Certain it is that Tillyer lived on to the age of eighty-six and old colleagues attended his funeral.

The B.B.C. continued to occupy the attention of the Board. There was scarcely a fortnightly meeting at which it was not mentioned. The newspapers were increasingly concerned about its growing importance. In January 1926 the number of current licences exceeded 1,840,000 – and this of course only covered the owners of receiving sets whose consciences compelled them to pay the 10s fee. An official report on the work of the Corporation up to the end of 1925 stated:

> These additional stations [Belfast, eleven two hundred–watt stations, and high-powered Daventry] have enormously increased the proportion of the population who can receive the broadcast programmes by means of a simple crystal set...
>
> The policy of simultaneous broadcasting by means of trunk telephone lines has been adopted and gradually extended so that selected items of the programmes emitted from any station can be relayed to any other station in Great Britain.
>
> 'Outside broadcasts,' i.e., the direct transmission of such items as speeches, plays, concerts etc., as distinct from performances at the studios, have been instituted.
>
> At the same time the B.B.C. have carried out a wide range of experimental work with technical apparatus.

The newspapers looked with grave suspicion at this insidious growth. They would not even discuss the broadcasting of football matches and horse races by means of running commentaries. But the General Strike of 1926 showed them that they could not merely stonewall. The strike virtually closed down the national papers while the B.B.C. broadcast five times a day. In such a crisis the country depended upon it for its news.

Gradually and cleverly, in the hands of Sir John Reith, the B.B.C. exploited its success. It provided its listeners with charts of field or course divided into numbered squares, and in relation to this key described the game or race. (This was the origin of the still popular phrase, 'Back to square one'.) And in 1927 it gave a very successful running commentary of the university boat race. This, people discovered, was not journalism. It was something quite new. Not until October 1930 did the B.B.C. 'steal' the first unquestionable news item, the crash

M

of the airship R101. Reith had chosen the right occasion to break out of the strait-jacket of restrictions, for the public felt it was only proper that they should have been given news of this disaster at once. It was a Sunday morning.

One cannot help sympathizing with the newspapers, but it is harder to understand the continued preoccupation of the agencies. Certainly the Exchange Telegraph had long been proud of its instantaneous reporting of the boat race by means of a wire laid along the river by a crew of linesmen temporarily turned sailors (a hilarious operation, one gathers). But, as unbiased people began to realize, there was room for both the spoken and the printed account. In any case the agencies were getting something for very little work. They collected the news in any case. Reuters, the senior agency, edited the bulletin. And in due course each – Reuters, Press Association, Exchange Telegraph, and the Central News until this last dropped out – got its share. At this early period the E.T.C.'s share was between £600 and £1,000 a quarter. Twenty years earlier, when three or four thousand a year had represented the difference between a profit and a loss, a dividend or no dividend, such a windfall would have been gratefully received. Even in 1930, though not riches, it was more than pin-money. And at this time there began the broadcasting of late market reports for which the Company is still responsible.

One small but specific complaint about a result of broadcasting appears in the engineer's reports. The aerials which were going up on the roofs of many houses were often makeshift contrivances. They had a way of falling down and damaging the Company's overhead wires.

There were many signs of progress and enlargement. In 1924 an office was started in Belfast, and within a month – from contracts and subscribers – was earning at the rate of £5,000 a year. Another storey was built on top of 64 Cannon Street, the object being to house the entire London communications system under one roof, the Panton Street premises, partly sub-let, being kept as a stand-by. The question of Ancient Lights, which arose in Cannon Street, was settled both amicably and simply: the owners of the building which found its view obscured were told that as far as the Company was concerned they were at liberty to build another storey also.

The 'Paris Office' had been the private apartment of André Glarner. In 1927 he obtained two rooms in the premises of a journal. The story that this was *La Vie Parisienne* is not true: it was *Le Petit Parisien,* which as part of the contract allowed free use of its news services.

In the records a few statistics catch the eye, indicating progress. In December 1925 there were 2,222 instruments at work. Stock Exchange calls approached 20,000 in a day. And Statistics, a department of which more will be heard, had acquired over seven hundred subscribers. The use of annunciators was not to survive to the present day in the House of Commons, or County Hall, but at this time with the requirements of the L.C.C. there were twenty-eight in use. Linesman Percy Davies tells a story about servicing the House of Commons annunciator lines at this period. He and his mate were working in one of the ventilating shafts under the flooring. Both felt that it must be about dinner-time but neither had a watch, so Davies crawled back and pushed up his head through a trap door concealed by a seat in a corridor. An M.P. and his lady guest were passing. 'What's the time, please?' Davies asked. The lady looked down at what appeared to be a severed head resting on the seat, and collapsed into the arms of her escort.

In the late 1920s the Board was much concerned over changes in the Betting Law. In 1927 Winston Churchill as Chancellor of the Exchequer put a tax on betting – as opposed to bookmakers who were already taxed. Some of the newspaper material to which the Directors' attention was drawn by the Chairman has rewarded research, at least in so far as it discloses some remarkably muddled thinking. The Jockey Club, rightly or wrongly, were clear enough: the tax would adversely affect the breeding of bloodstock in this country. In the House of Commons, particularly among the Labour opposition, the main concern appeared to be that betting had grown to such proportions as to have become a vice as serious as drinking. But by taxing it one would legally recognize it, which would be morally wrong. On the other hand it would be nice to have the money, as it was with the duty on drink. Mr Snowden rose to 'confess', apparently in all seriousness, that he had once put sixpence on a horse and lost. But he objected to the whole 'industry' of bookmaking and to the tax because it would encourage betting. One is reminded of the admiral who was shot *'pour encourager les autres'*.

In 1929 the totalisator appeared, first at the racecourses of Newmarket and Carlisle. The Exchange Telegraph had the practical interest that their subscribers, the bookmakers, were affected. (They preferred to call them subscribers, not book-makers.) The Company might also have felt a sentimental interest because, although they do not seem to have realized it, the tote was a twin, having been invented in 1872. It took fifty-seven years to reach this country and was only then available, so it was said, because the bookmakers had begun to fear they might be banned, and therefore might go on strike; so an alternative method of betting had to be provided to prevent betting being driven underground.

Remarkably, the bookmakers themselves seem to have been scarcely more concerned about this Trojan horse than they were about any other outsider. But within ten years of its appearance the turnover of the tote rose to £9 million a year.

Within the Company the most interesting developments concerned the welfare of the staff. The rented playing-field already referred to gave place to a sports ground at Wembley, owned jointly with the Press Association. A Sports Club was formed. When the essential equipment had been provided and a pavilion built, largely through a personal gift by Sir Charles Hyde, Chairman of the P.A., the Board provided only a small annual subsidy. The Club stood on its own feet. There were athletics as well as football and cricket. The Sports Club was a great success.

Individual credit for this is invariably given to Charles Wills. As Manager with an autocratic Managing Director-cum-Chairman over him he might easily have developed into no more than a yes-man. Instead he is one of the outstanding characters in the Company's history. He served in every non-technical department, but it was as sports reporter that he excelled. He was a sportsman both in work and play. It was said that he could write brilliant eye-witness accounts of half a dozen matches which took place simultaneously. His 'eyes' were the brief notes phoned in by the reporters on the fields – telling only who had scored and who won – plus his trained imagination. He would describe how so-and-so raced down the touchline and centred with his inimitable skill for someone else (whose style Wills knew) to put in a well-timed shot. He sat dictating to two shorthand typists alternately, each girl covering two or three matches. It was thrilling to read, and there is no record that he was ever caught out. Even if the telephoning failed he could produce an account.

Wills spent all his Saturday afternoons in sports reporting. Football he took in his stride, but cricket was his true love. He knew all the famous sporting personalities, always sent a telegram of congratulations to anyone who scored a century, and was responsible for bringing many cricketers into the Company. It was he who somehow acquired the Exchange Telegraph's reporting box at Lord's which looks onto the ground through a window next to the score board. Backed by the administrative skill of the Director Stanley Christopherson, his enthusiasm kept the Sports Club not only viable but vital. He was a performer too. He had captained the Polytechnic's 1st XI and he led out the Company's side from 1910 to 1938. Bald, bespectacled, and as serious-looking as the actor Robertson Hare, he had the happy demon of sport in his heart. There exists a photograph album of Sports Club activities – ladies racing with eggs in spoons, laughing or biting their tongues; ladies wheel-barrowing men; men, more serious as is their nature in the serious business of sport, triumphantly smiling or earnestly intent on winning a race or a match; there is happiness or absorption on every face. It is not too much to say that many of the Company's professional victories – which depended on team work – were won on the playing-fields at Wembley.

A parallel social affair was the annual Staff Dinner. Most office workers of a certain age have experienced such an occasion, and in the essentials they cannot vary much. One needs only to mention the data to wake memory or imagination. Three or four hours are spent in a scene and at a meal far above the standard of living of all but a very few of the guests. In uncomfortably formal dress, in the splendour of the Savoy Hotel or the immense Connaught Rooms, a couple of hundred people find their places from the table plan and eat five or six courses with a wine for each, listen to speeches, then begin to relax and possibly to enjoy themselves. Some may grumble at having to obey this 'royal command', but all would be offended if they had not received it. The Company's Dinners were always held on a Monday in February when there was no racing or other sport and when therefore most people were free. The financial account of one dinner happens to be available. The money in itself does not matter, but it accentuates a point. Two hundred and seven guests dined at a cost of £324 which included £106, roughly a third, spent on bringing the provincial staff to London. This was the only occasion when most of the London staff met most of their provincial colleagues. But the numbers in this instance are small. Later on the Chairman shook hands with up to six hundred guests.

Nobody remembers speeches and none of those at the Company's Annual Dinner were recorded. But it is striking that many of the senior staff remember what the Chairman's speech in 1935 was about. This was the year when Wilfred King received a knighthood in the New Year Honours. He began, 'You will want to know what the King said.' He seemed to be passing on a message. As a commanding officer is given the D.S.O. for the achievement of his battalion, so the Chairman felt he had been given the K.B.E. for what the staff had done. Wilfred King did not have the reputation of being a good after-dinner speaker. But consciously or naïvely he made his point.

It is sad that both the Wembley Sports Club and the Staff Dinner are things of the past. After the sale of the Wembley ground, cricket and football continued until recently to be played on rented grounds. Today's club activities include ladies' netball in Lincoln's Inn Fields, table tennis at St Bride's and Bishopsgate Institutes, and darts competitions. But the sports club as it used to be, with its own clubhouse and ground, is no more. It is so with other companies. Perhaps home life developed with television (though it lost in other ways) and men found other recreations like membership of the local golf club. Perhaps the difficult struggle homewards in the blitz began it, or new wives made new demands. We are not concerned with social history. But at this same period something which comes under the unsatisfactory heading of 'welfare' was born, developed, and is prospering today. As such it deserves to open the next chapter.

# CHAPTER 10

# *The Lean Thirties*

A MINUTE of the Board Meeting held on 26 June 1924, runs:

> Pension Scheme. Report from M.D. on interview with Mr Bisgood, late Secretary of the Edinburgh Life Assurance Company. Read letters from Mr Bisgood and the Board agreed to pay him a consulting fee of 100 guineas, and it was arranged that the M.D. shall ask him to submit the necessary scheme.

The 'late Secretary' gives no further sign of life so far as the existing records are concerned. But the date on which the Managing Director made this report is of interest. More than three years later, on 28 September 1927, the Chairman – who was of course the same person – reported to the shareholders: 'The establishment of a Pension Fund for the Staff is under consideration.' One may believe that the period between had not been empty, for it is elsewhere stated under this same date that the matter 'had occupied the careful attention of the Directors for some considerable time'.

A Fellow of the Institute of Actuaries, Owen Kentish, was brought in. He calculated that a sum of £26,800 would be initially required. In March 1929 a Trust Deed was signed, Wilfred King and General Anderson being the trustees representing the Directors and Charles Wills that of the contributory members of the staff. A transfer was made from the Company's reserves, of £27,000 worth of five per cent War Stock. The male members of the staff between the ages of twenty-one and fifty-five received a circular explaining the scheme, and were invited to join.

Wills, of course, could not be confirmed as trustee for the contributory members until there were contributory members to do so. The initial membership was two hundred and thirteen. Of these, two hundred and three voted for Wills and ten abstained. Abstaining suggests uncertainty. The Manager was extremely popular. One has heard him spoken of as 'loved by one and all'. He did not hide his kindliness under a bushel, or a beard, as did the Managing Director. Anyone could go and see him in his office at any time on any subject.

It seems possible that a few of the staff were shy about committing themselves in alarming legal terms to something they had no experience of. They may have hesitated 'to give to the Directors of the Company irrevocable authority to deduct from my monthly salary or weekly wage, as the case may be, my contributions as laid down by such rules.' Even the most kindly manager must be on the side of the Establishment. If the scheme was really in one's favour why tie one up like that, with a witnessed signature? But all the staff did not think in this way; perhaps only the ten who abstained from voting.

The Company's auditors acted as honorary auditors to the Pension Fund. The Company's Secretary, at this time W. C. Stevens, also gave his services. Within a year membership rose to two hundred and sixty-eight, and with it the capital. So the Pension Fund was born, and advanced.

Eric Steadman began work as chief engineer on 1 January 1930, and a number of changes of apparatus or technique followed in the next nine years. The new fast special sporting service was started, equipped with Creed page printers. These had nothing of the Victorian appearance of Higgins's machines. They were as sleek almost as the modern television set. At first they recorded at sixty-one words a minute, this speed being soon whipped up to sixty-six. Gradually the Creeds superseded their predecessors on all the services: in 1934 the Kleinschmidts, which had been on the Parliamentary service for almost ten years, and in 1939 the Law Courts instruments. So volubly did these new machines talk that they set up a record of over 2,000 million words in a year. To appreciate this Higgins-style statistic, it helps to know that if you count at the rate of one a second it takes between eleven and twelve twenty-four-hour days to reach a million. So to reach 2,000 million would take between sixty-three and sixty-four years.

A visible memorial to Higgins was removed when the oak derrick on the roof of Bartholomew House was taken down. For forty years it had carried as much as two thirty-wire cables, and two fifteen-wire cables plus numerous open wires. It had withstood wind and snow while other supports, approved by the L.C.C. Inspectors because they had the statutory number of stays, went tumbling.

In London many extensions of undergound piped wires were laid. The east-west system from the Monument to Hyde Park Corner was completed, as was the alternative route from Cannon Street via Fleet Street to Panton Street. In the provinces, as the local relay stations increased in number, more and more subscribers' instruments were connected up with them. And there was another change. These instruments were for the first time powered by the house electricity instead of by battery. It seems surprising that this innovation came as late as 1933.

About a year before the war an entirely new instrument was brought over from America to be tried out on the Company's systems. This was the Trans Lux. As it were, it read a tape and momentarily showed the words and figures in enlarged form on its translucent face. The advantage claimed was that a number of people could watch it from the comfort of arm-chairs, say in clubs. The disadvantage – soon pointed out – was that one had to watch attentively, for the information which appeared immediately disappeared for good. The Trans Lux was intended primarily for racing results and share prices. The Stock Exchange, famous for a type of wit, composed a poem on it. One verse is printable:

Trans Lux, on thy glassy screen
Tell us what the divs. have been,
Tell the House from end to end
What the Boards do recommend.

During the thirties there were two changes on the Board. In 1933 E. C. Barker died. He had become a Director on the death of Colonel Sheppee, King's predecessor as Chairman, in 1913. Barker was replaced by F. E. Davies, son of Captain Davies, and he died a year later in 1934. Francis E. Sheppee, son of the late Chairman, filled the vacancy thus caused.

In contrast to this comparatively early mortality, the staff continued to show that longevity and faithfulness may go hand in hand. C. Waller died after forty-two years' service, reporter W. J. Parker after forty-eight, operator Burrage after forty-five. Long service was not always terminated by death, although the more modest members of the staff did not as a rule enjoy their pensions long.

In 1933 Commissionaire Sergeant Williams retired, after thirty-three years' service, which is relatively short, but sergeants unlike other members of the Staff could not join as boys or young men. He is still remembered. He was a big man with waxed moustaches and the military bearing of the famous Kitchener poster, 'Your country needs YOU'. He was in charge of the messengers. With these restless, quick-witted boys he was a martinet – as was necessary. But he was also something of a father figure, protecting them from exploitation. If a member of the staff whom Sergeant Williams did not like or trust wanted a messenger, he might say, in a parade ground voice, 'Sorry sir, no boys available'. Sometimes the boys would chaff him guardedly and call him Sarge. When in a good humour he permitted this. But if they went too far he would snap into the disciplinarian again and they would retire to their corner with frightened eyes. With Wilfred King he enjoyed the intimacy of a regimental sergeant-major with the C.O. He would knock smartly, stride into the Boardroom and halt with a click of the heels to make some announcement. And King would say, 'Thank you, Sergeant,' or 'Do this or that, Sergeant, please,' almost as if to an equal.

W. J. Buxton retired in 1936 at the age of seventy-nine after fifty-three years' service. He lived only one more year. There is no record of when linesman D. Longhurst died, but in this same year, at the age of seventy-five, he 'felt obliged to give up work'. Those who have seen how a linesman earns his daily bread will be amazed he kept it up so long. But another linesman, George Piper, died at work a month before his seventy-eighth birthday. We are told by the statisticians that people are living longer and longer, but if there are now any centenarians on the staff they have kept it quiet.

The offices of Secretary and Assistant-Secretary seem also to have some affinity with the well of eternal youth. The first incumbent, Captain W. H. Davies, lived in command of all his excellent faculties to a ripe age. Wilfred King followed him as Secretary and would appear to have gained another span of life by the appointment. The next Secretary, ignoring a short interlude, was George Hamilton. It will be remembered that he was retired on the grounds of age in 1926. He lived until 1954 when he was eighty-eight, enjoying a pension from the Company. As an indication that he also retained an alert mind he continued for many years to earn a few hundred a year from his twenty-five per cent commission on the financial advertising he brought in from his wide circle of friends.

There are still a few active members of the staff who remember the Assistant-Secretary, Alfred Jackson. 'Pop' was popular, although almost the only specific things recalled about him are that he wore a tail-coat and was so meticulously economical he would at the end of a day search the office floor for fallen pins. It is said that some mischievous people used to drop pins to keep him occupied. He had joined the Company in 1882, and he retired in 1926. The Chairman then gave him a lunch at the Gresham Club. He also received silver plate and a pension of £400 a year. He lived another twenty-seven years to die at eighty-nine. All who have been Secretary since his day are still enjoying good health, and are also honoured.

In this connection the sixty-fourth General Meeting, which was held in September 1935, deserves mention. Those present are given as Sir Wilfred King, Chairman, Lord Sandwich, Stanley Christopherson, General Anderson, Charles Wills, Manager, William C. Stevens, Secretary; and as 'also present' Leonard W. Crouch, Solicitor, A. Jackson; and J. Kilpatrick and T. F. Watson representing the auditors. The meeting did no more than a well-conducted meeting should – it unanimously resolved that Directors' Report and Statement of Accounts be adopted, unanimously resolved that a final dividend of ten per cent on the 'A' shares and 3s 4d per share on the 'B' shares be paid. Alfred Jackson, as he had no doubt emerged from retirement to do, proposed that the Auditors, Deloitte, Plender, Griffiths & Co., be re-elected at the remuneration of £250 for the coming financial year.

The interest lies in the list of those present, for besides Wilfred King it includes the names of three future Chairmen. Since the last two, Sheppee and King, had held the position respectively for fifteen years and twenty-two not out, this can scarcely have been expected.

It is history that the 1930s were lean years. There were three million unemployed. Any young man who got a job with the Exchange Telegraph (or for that matter anywhere else) stuck to it like a limpet to a rock. But whether or not the salaries were aligned to the rapidly rising cost of living, the junior wages staff were still getting a pound a week or less. In the minutes there are fairly frequent references to rises of salary of £50 or even £100 a year. But as regards wages one reads: 'On the recommendation of the Managing Director increases of weekly wages amounting to [say] £2 10s a week were granted.' One is never told how many individuals were covered, and £2 10s is above the average. Once it was as little as 7s 6d, and to be shared by at least two, for 'increases' was in the plural.

Talking to members of the staff who date from those days, and are not now starving, one learns something of the personal anxiety and strain which lay behind those bald statements. No application for an increase would even be forwarded by the head of the applicant's department until he had been with the Company for several years, and once an application had been granted no other could be made until a year to the very day had gone by. Requests had to be put in on a Thursday before a particular Board Meeting once a month, so if the anniversary of a rise fell just beyond this Thursday the would-be applicant had to wait at least a month. Hard up as they were, they watched the calendar with hungry eyes. Applications had to be made on a certain blue form and in a stipulated manner, beginning, 'I beg to apply . . . ' and ending 'Your obedient servant.' At the bottom of the form the head of department would add notes on the man's punctuality and standard of work before forwarding it to the Managing Director. This was where the application might stick or be thrown back. If the Managing Director passed it, obtaining the Board's approval was automatic.

The next move was that the applicant was called into the Presence. This is not in the context an inappropriate word. One pensioner writing of Wilfred King uses a capital H for the pronouns He and Him, and it is clear from the form of letter that this is done naturally, certainly not as a joke. It needs little imagination to picture a young employee to whom 3s 6d more a week would mean a very real difference standing before the desk of the Managing Director with his frock coat, butterfly collar, imposing beard and whiskers.

Jack Ball, now the Property Manager, describes a specimen interview of those pre-war days. There was what seemed a very long walk from the door to the desk where the Managing Director sat absorbed in his work. After uncomfortable seconds or even minutes King looked up and said in a sharply questioning manner, 'Well, boy?'

At the first interview on joining the Company a boy was called by his name. Thereafter he might remain 'boy' until he reached executive level, when he became Mr So-and-so.

'Well, boy, you are asking for an increase to your wages?'

'Yes, sir.'

'How long have you been here?'

'Five years, sir.'

'And you are asking for an extra 3s 6d a week?'

'Yes, sir.'

The Managing Director then catalogued all the advantages accruing from being a member of the Exchange Telegraph. He concluded, 'Do you still ask for another 3s 6d a week?'

'Yes, sir.'

'Why?'

'I – I need it, sir.'

The applicant might have anticipated the course of the interview so far, but then came the typical unexpected question.

'What do you do on Saturday afternoons?'

'I, er – '

'Do you work?'

'No, sir. It is my afternoon off.'

'Then go and see Mr Coombes in the Telephone Room.'

The Telephone Room was in effect the sports news room. There on a Saturday the results of games and greyhound races were received from men who had spent the afternoon reporting. Thus a junior could earn an extra 3s 6d.

Ron Witte, Telephone and Transport Manager, tells of a similar interview when he was in age a boy. He was so overcome that he was tongue-tied and began to tremble violently. King stared at him silently and then himself began to tremble. And so for some minutes they remained. Witte got his rise.

Eric Martin never got over his boyish awe. When he became a senior member of the staff he might have occasion to see King several times in a day, but he always, instinctively, paused in front of the closed door to smooth his hair, straighten his tie and pull his coat straight before he knocked. He had joined the Company when he was seventeen as a junior clerk in the general office at 7s 6d a week, and at this moment he continued to behave as he had while in his teens.

It must not be supposed from what has been said that King was ever rude or bullying. His manners were faultless. That, in part at least, was what was so awe-inspiring. He always said 'thank you' for good work done. But his measure of good work was exacting. Once James Fullex, when he had just covered the whole of the Wimbledon fortnight, reported elatedly to Wilfred King that the Company had been ahead of its rivals in giving the result of every single game except one. 'Why were we behind with that?' King asked.

One more reminiscence of the hungry years. Two linesmen were sent to install an instrument in the offices of Tate and Lyle. They did the job quickly and efficiently. The office manager said to his secretary, 'Give them each two pounds.' Two pounds, more than a whole week's wages! They followed the secretary with beating hearts. They were each given two pounds of sugar!

The personal agony of asking for a rise ended when the trade unions entered the scene in the middle thirties. They had a difficult entrance. The National Union of Press Telegraphists was the first. Their representative asked men to join, promising almost to double their wages. But no one dared from fear of being sacked in those lean days. 'If your Managing Director accepted the Union, would you join?' was the next question, to which the seemingly academic answer was 'yes'. The Union officials requested an interview with King. He refused. They sent in the message that it would be to his advantage to see them.

King saw them and said bluntly that any member of the staff who joined a union would be dismissed. 'In that case,' said the Union officials, 'no news from the Exchange Telegraph will be handled on Fleet Street.'

That broke down the barrier, and those who joined *did* have their wages almost doubled. But the new arrangement was not one-sided in those early days. The Union saw to it that their members were punctual and hard-working, that they kept their side of the bargain. Otherwise they were thrown out. Members were more afraid of the Father of the Chapel than of their departmental boss.

Lean years though these were, the Company as a whole did well. At the beginning of 1930 the estimated rate of revenue was £280,000. By September 1938 it had risen to the then record figure of £380,000. After that it dropped a little, but not much. There was also £92,000 invested and £15,000 in the deposit account. One is inclined to ask how it was that the Company did as well as this and why – in the opinion of the staff at least – was it so mean about wages. The second question is quickly answered, and not only by the stock phrase that no staff is ever satisfied with its remuneration. The Exchange Telegraph's pay *was* low by Fleet Street standards. This was because the Managing Director had made it his business to know where every one of the Company's pennies had come from and where every penny went. Like someone from a poor home, he had early learned the habit of economy, and he maintained it compulsively or from a sense of duty. Only with his own money did he feel at liberty to do good by stealth.

The question how the Company managed to do as well as it did in this period is more difficult to answer. On careful consideration (as the Board would have put it) this was because in place of satisfaction there was a sense of restlessness and urgency to push on, such as you feel (perhaps uncomfortably) when you mount a good horse. Certainly there was also, paradoxically or not, great care over detail. The minutes show that the smallest question was thrashed out. They also show that many risks were taken by this careful crew, as when they poured money into the Central News which almost as regularly lost it until it was finally taken over. Minutes, to digress, are fascinating if given

'careful consideration'. It remains for someone to write a novel
in this idiom.

Although not going quite as far as that, we will try to picture
the men in authority at a Board Meeting in, say, 1935, the
imagination being prompted by photographic portraits and by
discriptive fragments. The Chairman is kingly in more senses
than one. He holds himself upright in his chair. Moustache and
beard are neatly trimmed, and one can just make out the
slightly indented ring left by a top hat on his still amply covered
scalp. He glances round the gathering with eyes which are
slightly protuberant, as if from the habit of looking closely into
all things. But the neck enclosed by the stiff collar is thin and
wrinkled, and the slightly claw-like hands betray his age. Near
him sits Charles Wills, the Manager, the nearest thing to a
confidant that W.K. ever had. He is fifty-five. He is bald, with a
long nose, a small moustache and steeply arched black eye-
brows. He has the appearance of a suddently awakened owl,
except that the wide eyes suggest mild amusement in what they
see – a very kindly person, in other words. There is Brigadier-
General Anderson, fifty-five years old, with facial features
drooping somewhat a little like a bulldog, but with straight,
clear eyes which are a little sad. A dignified and charming old
soldier, impressive and *simpatico*. General Anderson is a personality in his own right. He served with distinction in both the
Boer War and the First Great War. Besides the D.S.O. he won
the Legion of Honour and the American D.S.M. He has given
what might have been his retirement to the business his ancestor
founded, and has already been learning the trade for seven years.
But he is not by nature impatient. He knows now that that man
at the head of the table will never retire. But he is committed as
he once committed units to battle. There is no drawing back
and he does not want to. He has become interested – in the
Joint Committee, in the provincial tours he shares with Wills, in
dealing with the trade unions, in talks with the B.B.C., in
representative journeys as far afield as South Africa. The Chairman gradually passes over more and more to him.

There are four other Directors. One is Lord Sandwich, clean-
shaven, distinguished in appearance, relaxed as a man who is
perfectly sure of himself. Another is Stanley Christopherson, a

N

well-made man and handsome, spruce and alert, well dressed
in city clothes of a much more modern fashion than those worn
by the Chairman. A shrewd, successful businessman would be
one's guess. Francis Sheppee wears a tweed suit, looks healthy
and takes things calmly – he might be a country squire with
something in the City. The rest, the Secretary W. C. Stevens
and T. F. Watson the representative of the Auditors, are still
alive and therefore best left undescribed.

Among the subjects which particularly occupied the Board in
1935 and the immediately subsequent years was that hardy
annual the Joint Agreement. Not that the Joint Service was
still causing trouble – it was going remarkably smoothly and
profitably – but in conjunction with the Press Association a
number of the Central News and Column Printing Company's
shares were being bought.

The lease of twenty-two Budge Row was bought for £4,000
to provide a new home for the statistics department. This had
continued to expand. But Ridley who had been largely responsible for getting it off to a good start was ailing, and in September
1937 he retired with 100 guineas in his pocket 'as a token
of the Directors' esteem and regard'. (He was succeeded by
R. W. Brash.) New offices were also acquired in Galway and
at Salisbury, the latter at the remarkable rent of £15 a year
including heating. This, no doubt, was another example of
Wilfred King's economy. The provincial offices were generally
in unfashionable streets, small, and on the top floor. The furniture looked as if it had been bought up in second-hand shops – as
it probably had. Quite often the roof leaked.

When the Stock Exchange was pleased with the service provided by the Company it expressed this by asking for something.
It now asked for thirteen more telephone boxes. One supposes
– although this is nowhere stated – that the telephones were not
still housed in the cloak-room.

But the most rewarding subject for research is horse racing –
whether one is interested in the sport as such or not. This is
because the betting laws and various other restrictions, and also
the need for speed, still made the reporter's work unconventional.
Very little of this got into the minutes, of course, but we may
start with something that did. In November 1936 the Board

(which we have tried to picture at its dignified deliberations) received the report from the Managing Director that two hundred and twenty-five Belfast bookmakers (more than one would have thought could exist in one town) had each been fined £5. Bookmakers, generally referred to as subscribers, were the life blood of the racing service, and at the time there was no explanation of this outrageous affair. Nor for eight months of Board Meetings do you find any. Then it comes out – for the persecution was still going on – that it is all the fault of the Orangemen. A Protestant anti-gambling conglomeration had found some damn-fool point of law which naturally had never been enforced and were seeing that it was. They were making speeches and all, exhorting the Government to take still sterner action. The Boardroom minute (somewhat expanded here from other sources) concludes sadly, 'There seems little prospect of any alleviation'.

One cannot discover how the matter ended. Charles Wills and George Wray paid a number of visits to Ireland and eventually sorted it out.

In England the bookmakers went free, at least from religious persecution, but the lot of the racing reporter was, although it sounds a happy one, not easy. If he enjoyed his work it was because no man worth the name ever ceases to be a boy, and the motto, Accuracy, Impartiality, Celerity, could only be achieved by the ingenuity of naughty youth. The starting odds, the 'Off', and the result, had to be passed to London immediately. How could that be done with no telephone on the course, or none available perhaps within miles? Earlier we told of a telephone at a hotel within visual distance of the course. But there were not often such amenities.

The symbol of the times was Alf Pepper's telescope. There must have been other means of bridging the difficulty – racing binoculars or a long string of watchers. But that yard-long concertina tube of brass and glass, a telescope such as the most daring skipper of an adventure story might have put to his eye, if he was strong enough to hold it up, is typical of those derring-do days. What the telescope had to do here was to read the tic-tac of the man on the course. Tic-tac to most people is still more of a mystery than Cockney rhyming slang, even than bush telegraph. But five minutes with a co-operative race reporter has revealed clearly enough to the un-race-minded writer the working principles of this communicatory Swedish drill. Rarely are there more than ten horses in a race, almost never more than twenty. They and therefore their jockeys are numbered on the racing card, which is issued in advance. All that people want to know is the betting and the finishing order of the first four. This can be conveyed entirely by numbers. Without giving away the code one may say that it depends on conveying a comparatively small scale of numbers by identifiable gesticulations of arms and hands.

So much for theory: now for practice. Very probably there was no telephone available even within the long visual range of the telescope. But probably, since the race course did have a telephone of its own, there was a line of telephone wires. So the team is increased from two to three – the man on the course, the man with the telescope (or racing binoculars), and the man who has climbed up a ladder and uses, by arrangement with the Post Office, a tapped wire.

How it worked out was this: the man with the telescope had plenty of time to read the odds from the tic-tac man on the course. He shouted these up to the man who had the telephone. The telescope man could also see the horses start, and called out 'They're off' to the man up the pole. After that he depended on the man on the course to give him the finishing order by tic-tac.

Of course it did not always go so smoothly. The man at the business end of the telescope was then Alf Pepper, one of the great back-stage boys. He is no longer available. In September 1944 he completed fifty years of service; two years later he reached retiring age but continued to work and he died at Lingfield races in 1954. His son Sam, who maintained the family tradition in the racing service for character and initiative, died in 1970.

Louis Nickolls who edits the Company's House Journal fairly recently unearthed a couple of Alf Pepper stories, partly from Jack Atherton who is now Manager of the Manchester office but was at the time of the incident 'a young chap on the engineering staff at Leeds'; and partly from 'Pep's' son, Sam

Pepper. Jack Atherton was sent to help Alf Pepper at the Pontefract races, and in his eagerness saw himself breaking into the
romantic field of race reporting. As it turned out his first job
was to carry the telescope and tripod from the railway station to
a point on the main Pontefract-Castleford road. There he
found a candlestick-type telephone placed on a board which
was nailed half way up a telephone pole, and connected in to
the line and the Leeds office. Pepper set up the tripod and
trained the telescope on the tic-tac man on the course. A third
man, who he thinks was Jim Smith, climbed a short ladder set
against the pole. From the top rung he could just reach the
telephone. Atherton's second job was to hold the ladder steady
and pass on the odds, the 'Off', and the result, while Pepper,
entirely absorbed in what he saw through the telescope, called
them out.

There is another story told by Sam Pepper of when his father
and Bill Lee were again covering a Pontefract meeting 'from
the outside position'. The story runs:

> Father had his eye glued to the telescope and was reading
> aloud the runners and jockeys to Bill Lee who was on the
> 'phone up the ladder. Usually Bill repeated whatever
> message was called out to him. But on this occasion there was
> no repeat.
>
> 'Have you got 'em, Bill?' my father called out.
>
> Still there was no reply. Looking round, my father saw the
> reason. Still clutching the telephone, poor Bill had fallen off
> the ladder into the ditch.

Unfortunately that is where the story ends. One would like
to know what the Managing Director said when he enquired
into the reason why the report of that particular race was
delayed.

The reporting just described was easy compared with that at
other courses. At Goodwood in the early days they had to use
carrier pigeons because the Duke would not allow telegraph
poles on his extensive lands. And at one small Irish course
which was on the coast and enclosed by hills the telescope man
had to use a boat and signal to a telephonist on the other side of
the obstructing hill.

Rival reporters, of course, were up to equally ingenious tricks. They had to be sabotaged. No methods were barred in this lively game. If wet blotting-paper is inserted in a telephone it will fuse it, and of course lines can be cut.

Every member of a news agency's staff, as of a newspaper, is a potential reporter. Len Alldis, the librarian, has an interesting reminiscence. He joined the Company as a messenger at 15s a week in 1926, and used to carry the Telegraphic News-sheet on the various runs in the area of the clubs. One Saturday evening thirteen years later, by then a junior member of the editorial staff, he was walking with his girl friend (now his wife) along Victoria Street. They arrived at Victoria Station just as an I.R.A. bomb went off and injured seven people. Alldis did not waste this heaven-sent opportunity, but immediately rang the office. Most of the papers took the story. It was made the more piquant by the coincidence that the House of Commons was at that moment debating the Prevention of Violence Act. Alldis still has the chit for one guinea which he received for this scoop.

As the 1930s drew towards their close the rumours of war increased, as did preparations for that eventuality. The engineers were busy adding to the number of alternative underground routes for the circuits, and arranging stand-by sources of electricity supply. In October 1938 the Managing Director reported that the A.R.P. arrangements would be completed at all the offices within the next few weeks. He also spoke about the effect of the European crisis on the Company's news service to and from the Continent. He read letters from A. E. Ruttle of the Berlin office and André Glarner from Paris. Interest in foreign affairs was naturally keen. But could the existing communications be maintained if the situation deteriorated much further? General Anderson told of conversations he had held at the Foreign Office, and a talk with sub-editor R. F. Roland who had just returned from a visit to Germany.

If war came it would be of particular importance that the neutral countries should receive news from Britain. In April 1939 an office was opened in Zurich, and negotiations were started with the Foreign Office with regard to arranging a news service to Portugal. The news agencies set up a war-time Information Committee on which General Anderson was the

Company's representative. Emergency centres were set up, at Tring for the statistics department and Dunstable for the engineers. That common phrase of early 1939, 'If the balloon goes up', changed to, 'When the balloon goes up'.

There was a Board Meeting on Thursday 14 September. When, half way through it, the item 'Licence with City of London Corporation in respect of Underground Pipes in Threadneedle Street' had been dealt with, the Managing Director is minuted as reporting 'that the Prime Minister announced on Sunday 3 September that Great Britain was at War with Germany as from eleven a.m. on that day'.

That was that. All preparations had already been made, so the meeting went on to discuss some remarks recently made by Mr Robbins of the P.A., negotiations with the B.B.C., and to examine the accounts of the Central News.

If Wilfred King had been born four centuries earlier he might have played bowls with Sir Francis Drake.

# CHAPTER II

# Hitler's Interruption

NO WAR has so much involved the civilians of this country as did the one which Hitler caused. Scarcely had Neville Chamberlain's sad, tired voice announced the beginning of hostilities than the sirens wailed. That came to nothing, but in midsummer 1940 air raids began in earnest, first at night, then by day. The Managing Director reported on an interview with Mr Webb of the Stock Exchange 'in regard to assisting Members who have been inconvenienced through bomb damage affecting the Bartholomew House system'. No serious harm was suffered, however, until the new year when, on 10 January, the Portsmouth office was burnt out. Eleven days later the Swansea relay office was completely destroyed. On 23 April the Plymouth office suffered the same fate.

The first phase of enemy attacks had been directed chiefly against ports and shipping. By night these were comparatively ineffective; by day the R.A.F. made them expensive. It was then that Goering began the indiscriminate attacks on London and other cities with the prime object of breaking civilian morale. In this, as is well known, the Luftwaffe failed. But they did a lot of material damage. Nos. 21 and 22 Budge Row were hit by fire bombs; the annunciator system in the House of Commons was put out of action, wires and apparatus in Liverpool were cut or destroyed.

A part of No. 62 Cannon Street was requisitioned, the head office staff being packed into the warren of No. 64. At this early stage of the war it escaped damage, but the staff did their stints of fire-watching and there were many broken nights. One man who was then a junior tells of a fire raid when the tide was low and there was therefore little water for the hoses.

From one which was temporarily lying idle there flowed a trickle. A bearded old gentleman shook a weary fireman by the shoulder and, pointing at 64 Cannon Street, said in an authoritative voice 'Play your hose on that building, my man.' The Chairman, who was watching over the Company's property in the midst of fire and flying metal, was over eighty years old.

After the initial quiet period in the line, 'the phoney war', casualties began to be reported from the fighting areas. G. P. Pepper of the Doncaster office was killed in action in the Middle East. There were others later, E. A. Coleman, R. O. Hill, S. C. Gunn among them. But in this connection the most striking feature of the war years is the number of the Company's old servants who died from natural causes either in harness or shortly after retiring. In February 1942 E. H. Longhurst died at the age of eighty, after five years' retirement. He had been a linesman for forty-three years. James Yeo, a Law Courts reporter, died at work aged seventy-three after forty years' service. In the summer of the same year Owen Smith died at seventy-four after being a reporter for twenty years. F. W. Saville and W. Sidebotham died after each had served for thirty-five years. In the following year T. G. Boyland, Law Courts reporter, died at sixty after thirty-four years with the Company. And T. Shears of the Telephone Room staff died suddenly. He was only fifty-three but had been with the Company 'for over forty years' – which means that he joined at twelve years old or less.

These facts once more underline the tradition of long service. They suggest something else as well, for by no means did all the senior staff die during the war. There were other oldish men at work all those years; some youths and young men certainly, but not much in between with the able-bodied getting into uniform. The Board was of men who were more than mature. Yet the Company did well during the war in very difficult conditions. Now when it is often said that one is old at forty it is well to point out that some have carried on successfully when nearer to forty multiplied by two.

It might have looked at the beginning of the war as if the E.T.C. – as the Company had come to call itself – was going to be short of work. No one was much interested in cricket or football. In October 1939 the Blower – here still meaning the London and Provincial Sporting News Agency – and Tote Investors Limited announced that they were suspending operations for the duration. All racing had been stopped on the declaration of war – until the Newmarket Meeting on 18 October. After the Windsor Meeting on 27 December it was again stopped (this time by the weather) until the Newbury Meeting at the end of February. For one reason or another racing was severely curtailed. And in June 1940 the Jockey Club announced that there would be no more until further notice. One cannot help feeling that Wilfred King must have been pleased. He could concentrate on the things that interested him.

The Stock Exchange service continued, and as we have seen soon suffered 'inconveniences' – worse than snowstorms. The general news service was of more importance than ever. The first task was to appoint war correspondents. J. N. G. Holman was attached to the British Army in France. The Paris correspondent André Glarner went to the French Army's sector of the front, R. F. Roland to the R.A.F.

Glarner's appointment as war correspondent, although brief, was fortunate. His reports were excellent. They make interesting reading even today – or particularly today, for they vividly convey the 'Maginot mind' of that first winter of 'phoney' war. As such they deserve to be quoted at some length. Glarner was not young. He had reported for the Company in the 1914–18 war.

10 October 1939

The Maginot line has often been described as a huge battleship perched upon a chain of hills. I visited it with a group of British correspondents – who were incidentally the first civilians ever to inspect this particular front. To my mind the line rather resembles a gigantic submarine just rising to the surface with its periscope and turrets beginning to peep out of the water . . .

I entered a filter room with its giant machines capable of providing the fort with fresh air for months even if it were completely surrounded by the enemy or poison gas. Electric power is supplied by the central power station for the line, but emergency plants in the fort itself can be put into operation quickly in case the connection with the main plant should break down or be cut off. Thus the garrison of the fort could withstand a siege indefinitely operating all its guns and making use of all its defences and offensive equipment. From the power room we went to the telephone and wireless room where communications with the outside world can be both received and sent. Near by was the operating room as well equipped as any private hospital . . .

In the course of our inspection an 'alarm' was given for our benefit. The men went to their stations smartly. The great turret – three hundred tons of refined steel – was put into operation by its special motor. Hydraulic pressure can also be used to operate it if the motor should break down. In less than fifty seconds the men were at their stations on the three upper floors and the turret was reported 'ready for action'.

The central part of the turret, a huge tube some eight feet in diameter, can be moved up or down as required. It can also be revolved thus enabling the gunners at the top to fire in any direction.

High powered automatic machine guns are also an important feature of the defences. They can spray a front two hundred yards wide with eight hundred bullets a minute. Then there are automatic grenade throwers, just in case the Germans should get within close range; a new model anti-tank gun, marvel of French engineering sciences. These guns are said to be capable of piercing the armour of any tank yet made, and putting it absolutely out of action. Outside the fort were other defences against tanks.

27 October 1939

The men know that their war materials are of the finest. They know as a result of recent fighting that military technique has changed since the last war . . . After six weeks of fighting in 1914, ten of France's richest Departments were in the hands of the enemy. Today on the other hand not one German soldier stands on French soil.

20 November 1939

The French frontier – a wall of steel and concrete with outer lines of defence and the mighty works of the Maginot Line backed by yet other walls of resistance – has been brought to a degree of strength hitherto unknown in military history. There now exists a second line of fortifications in addition to the Maginot Line, and in some places there is a third powerful line of defence. Since early September thousands of picks and shovels have been wielded . . .

Glarner adds touches of humour, as when describing the inspection of some young recruits by a general. He asked one, 'My boy, do you know what a hero is?' The youth, shaking in his boots, replied, 'Oui, mon Général. A hero has lots of guts and plenty of hair on the chest.'

'Splendid! Can you name me a hero?' the General asked.

'Yes,' the boy answered, 'Joan of Arc.'

In spite of all his praise of the Maginot Line, Glarner appears to have been slightly concerned about its limited extent. Certainly he passed on the official French military point of view. But at least he mentioned the possible weakness.

5 November 1939

. . . There are many who think that the Maginot Line is a defensive line running only from the Moselle to Basle. This is not true. Only today . . . I was able to visit a number of formidable concrete casements, gun emplacements and ground defences which guard against possible attacks across the Belgian frontier. These fortifications have been built to forestall any German offensive through the Low Countries.

Very early in the spring things started to warm up.

14 February 1940

The most striking feature along the Western Front during the past fortnight has been the break in the absolute calm which had prevailed along the Rhine for the last five months. Since the last days of January, indeed, there has been a significant exchange of Franco-German machine gun fire from the blockhouses lining either side of the Rhine . . .

One knows all too well how this greater activity developed. The impregnability of the Maginot Line was never put to the test, for Hitler outflanked it by violating the neutrality of the Low Countries. Glarner had little more time for general reporting. His last despatch is dated 23 May. He had not yet lost his confidence.

I have seen German tanks spitting fire through every slit and aircraft bringing devastation with incendiary bombs and machine guns. I have seen the whole German mechanised rush that is now moving at a slower pace than at the outset . . .

Now the Germans with their own new technique are still hammering at the Allied lines, but our hour to hammer back is soon coming.

On that same day the Managing Director had the melancholy satisfaction of reporting to the Board that the Company had been first to give the news of the German entry into Holland and Belgium – from Madame Roetener, of the Dutch army laying down its arms – from Van Houten, and of the German invasion of Denmark – from W. H. Kelland. Kelland was Manager of the Copenhagen office, and his subsequent fate was not known until the armistice. Glarner managed to escape to England about the time of the Dunkirk evacuation.

The story of the war for some time thereafter was of blood, sweat and tears. But there were a few inspiring incidents.

On 28 March 1942, J. N. G. Holman and Edward Gilling, as the Exchange Telegraph's correspondents, accompanied the raid on St Nazaire. It is worth recalling the main details of this singeing of Hitler's moustache. The Germans had highly organized the port as a submarine base for their attacks on convoys. It also possessed the only dry dock on the Atlantic seaboard large enough for the biggest German battleships. The *Bismarck* was making for St Nazaire when she was sunk in May 1941, and the *Tirpitz* would undoubtedly have gone there as a further menace to allied convoys.

The raid was a combined operation, the military force being made of picked men from the Commandos, mainly officers. The naval units consisted of the ancient HMS *Campbeltown,* half filled with concrete, three destroyers, a motor gunboat, a motor torpedo-boat and a number of motor-launches.

As a rule only one correspondent was permitted from any agency or newspaper on such raids. But on this occasion, since it was quite likely that the man might not come back, two were allowed to go from the Exchange Telegraph's staff. J. N. G. Holman was in the gunboat which led the raid, with Commander R. E. D. Ryder. Edward Gilling was at the other end of the line in a motor-launch.

Holman had already had quite an exciting time as war correspondent. He had gone into Belgium with the British forces which opposed the German *blitzkreig* in 1940, was almost cut off at Arras and finally got away from Boulogne under German machine-gun fire. But at St Nazaire he had what he describes as his stickiest experience.

The gunboat led the way in to the harbour and immediately came under very heavy fire. This temporarily ceased when a rating wearing German uniform exposed himself on deck. When firing started again the *Campbeltown*, as much the biggest vessel, drew most of it. She forged ahead, barged through the lock gates, and sank – to remain there in the way of every German vessel for most of the remainder of the war. The rest of the force carried out their parts in the raiding programme, and then those who survived and still had serviceable craft got away. The gunboat sank on the way back, Holman and the rest of her complement being picked up by a destroyer. The motor-launch with Gilling aboard was feared lost but turned up in a home port a couple of days later.

The St Nazaire raid was as daring and successful as that of Zeebrugge in the First War. The Army won two V.Cs., the Navy three, one going to Commander Ryder. At that time he only had the Polar Medal. Thus he became the first man to wear the white ribbon of the snow and the red 'For Valour'. Gilling and Holman were Mentioned in Despatches.

The Company's war correspondent Arthur Thorpe twice escaped death when the aircraft-carrier to which he was attached was sunk. On the first occasion the ship was the *Ark Royal,* on the second the *Eagle.* The following is his account of the *Eagle's* last moments. It appeared in *The Times* on 14 August 1942.

I was in an ante-room with three officers when soon after one p.m. two tremendous crashes shook me out of my chair. I knew what that meant and leaped for the door.

As we opened it two more violent explosions rocked the ship. The hiss of steam filled the air, and I saw clouds of it pouring up from below into the broad after-deck across which we were running. As we dashed through the bulkhead door to gain the upper deck the ship was heeling over, and the water was washing about our feet. We scrambled up the ladder to the upper deck with the ship listing over terrifyingly to the port side, on which we were.

The sea, normally ten feet below, was surging a bare two feet below the rails. We reached the quarterdeck, grabbing at any projection to haul ourselves up the steeply sloping deck to the starboard side, and, clutching the bullet-proof casing enclosing the quarter-deck, I found myself next to the First Lieutenant, who was blowing up his lifebelt. I followed suit.

Looking round I saw the deck slanting more sharply than a gabled roof. Six-inch shells, weighing a hundred lbs., tore loose from their brackets and bumped down the deck. Ratings on the port side saw them coming, and flung themselves over the side to escape injury.

Turning to Number One, I put the quite unnecessary question: 'Is she going?' His answer was a nod. Several ratings who were hauling themselves up the casing clambered towards us; they made fast a stout rope and slithered down into the thick oil which was welling out from under the ship and coating the sea. With a confidence in my lifebelt which now amazes me, I slid down after them and let go.

I went under the water, and when I came to the surface I realized with horror that I had not properly inflated my lifebelt. My head was barely above the sea, as, with all the poor swimmer's dread of deep water, I splashed and kicked clear of the ship. As I worked my way out of the oil patch the water was more broken, and every short wave washed clean over my head until I was dizzy. No wreckage to which I might cling was within reach, and I confess I gave myself up for lost. Then, as a wave bigger than its fellows lifted me up, I saw the glorious sight of a cork float net twenty yards away, with sailormen clinging around it.

I fought madly to reach that float. Three times my head went under, and then I saw the float net again, this time only a few feet away. I made a despairing snatch and missed, but with another wild clutch I felt my fingers lay hold.

Half a dozen ratings clinging to the net tried to loosen one of the ropes and open out the raft which was tied up in a round bundle. Their fingers, like mine, were coated with oil and next to useless. The water was quite warm, but my great anxiety was the difficulty of holding on with my oil-smothered hands. Then another rating swam up and caught hold. He told us his leg was broken, and we helped him to crawl into the centre of the bundle.

The waves were breaking over us, and I hauled myself up to look at the *Eagle,* two hundred yards away. She was lying on her side and, down the great red expanse of her underside, men, swarming like ants against her great bulk, were sliding down into the sea. Then, suddenly, I felt a shock to the base of my spine. We knew it was a depth charge from a destroyer hunting the U-boat which had attacked us. Six or seven times this curious shock from below the waters shook us on the float net.

'She's going!' one of my companions gasped, and there was a mighty rumbling as the sea poured hungrily into the stricken vessel, forcing the air out of her. Water threshed around and above her in a fury of white foam, and then subsided.

The *Eagle* was gone.

After a moment's amazement at the sight we looked around hopefully, and cheered when we saw a destroyer only a hundred yards away, and making for us. We were soon alongside, and ropes, nets, and cork lifebelts snaked down the destroyer's side. Smiling faces from the destroyer encouraged us as my sailor companions seized the ropes and hauled themselves aboard. I clutched a trailing rope, but could get no grip with my oily hands until a rating already half-way up slipped down to me, and I used his legs to get a purchase.

Feeling battered and as weak as a kitten, I managed to slip the bight of a rope under my shoulders. Just then a wooden ladder clattered down the destroyer's side, and I succeeded in dragging myself up and aboard with a helping pull on my shoulders, feeling half dead and looking as brown as a nigger from head to foot with fuel oil. The decks were crowded with survivors, and scores more were being pulled aboard.

We raided the bathroom and rid ourselves of the worst of the oil. The destroyer's officers and crew were wonderful. They opened up their kits and their stores, and soon we were all equipped with dry clothing. Many of us were sick through having swallowed liberal quantities of oil, but tots of rum put queasy stomachs right, and soon we were laughing and joking at our quaint costumes. Some officers were dressed in long pants and vests, others wore football jerseys, grey flannels, and coloured shirts. Some had found shirts, but from the waist downwards were clad in towels draped around the waist.

Our late ship's war cry, 'Up the eagle's!' rang out as Captain Mackintosh came alongside on a float net. He had held his command for a bare six weeks. Then I saw an unforgettable scene. Another ship drew alongside, her decks packed with men from the lost carrier. As officers and men in the destroyer recognized those aboard her there was a bedlam of glad cries of recognition and bantering cheers.

On another encounter with the enemy Thorpe's luck failed him. Shortly after D-Day, being at Portsmouth and bored by inactivity, he went out on a naval patrol. They met hostile craft which fired a single burst before making off. Arthur Thorpe was hit in the head and killed instantly.

At home, a great deal of work was put in on the Continental news service. In March 1940 General Anderson went to Paris and conferred with André Glarner, Garrett of Zurich and Bordallo Pinheiro of Lisbon. From these three cities most of

o

the British news was disseminated. But Rumania was also interested, and in conjunction with the Ministry of Information the service was extended to the Balkans. The Lisbon office, although it continued work throughout the war, appears to have faltered somewhat. It began well. Pinheiro cabled his satisfaction with the news supplied to him, and the M.O.I. paid the Company £1,500 for the service in 1940. But after that they suggested that the bulletin should be reduced, and as from the end of 1943 they withdrew financial support for the service to Portugal and its colonies.

Conversely the Zurich office prospered exceedingly. There is, for instance, a record of particular interest shown in Switzerland in the despatches about the invasion of Sicily in the summer of 1943. This was natural enough, for it was the beginning of the attack on what Churchill called the soft underbelly of Europe. It proved not so soft after all, as those who served in Italy know well. But at last the Allied Forces were in Europe to stay, and advancing straight – if slowly – towards Switzerland. That small country was surrounded by war and probably only escaped the fate of Denmark and the Low Countries because with its highly trained mountain troops it would have been a hard nut for Hitler to crack.

Portugal, on the other hand, was not even on the touchline. When the fear that Hitler might invade Spain to get at Gibraltar from the rear had gone Portugal could relax as far as possible involvement was concerned. As for Rumania and the Balkan countries, they became too much involved.

Before leaving this subject it must be mentioned that early in 1944 Basil Gingell, then the Company's correspondent with the Fifth Army, had a lively time. In September 1943 he was slightly wounded when his three companions, all war correspondents, were killed. Later he was mentioned in despatches for 'gallant and distinguished service'. Finally, he covered the year-long Nuremberg trials and was one of the two British correspondents present at the hangings. The moral demands on a war correspondent are perhaps insufficiently realized. A soldier in the line runs the risk of being killed, but the fact that he can also kill is a psychological comfort. The war correspondent risks his life by the equivalent of going to the office. How many of us, however keen on work, would go to the office if we were likely to be shot? A commanding officer can only hope to obtain a small quota of medals and mentions for the men under his command. He is naturally inclined to restrict his recommendations to his fighting men. For an outsider like a journalist to be mentioned means that he has done something – or more probably a lot of things – that could not be ignored.

On the home front the engineers probably had the hardest time, repairing bomb damage and keeping the circuits working. Air raid and flying bomb experiences were a bane of those years, and no more will be repeated here – except to say that the annunciators suffered casualties. 'Four destroyed by enemy action,' stated one engineer's report. Bill Bain set the type for the House of Commons annunciator all through the war, singlehanded. He is the hero of an act which, if not in the medal-winning class, displayed a quite extraordinary sense of discipline and self-denial. And it has never been recorded. At one time the Commons met in the House of Lords, and of course the annunciators had to go with them. It was a make-shift arrangement, and Bain had to work in a narrow, airless room which housed the lighting switchgear. He was on duty for a very long time and he was suffering from 'flu. He would not give in and finally collapsed. A doctor was called for, and Lord Moran who was in the House – no doubt to watch over Winston Churchill – attended Bill Bain.

The distinguished doctor knelt beside the prostrate man, loosened his tie and collar. Then, probably diagnosing nothing worse than exhaustion, he pulled a flask from his pocket and held it to Bain's lips.

At the smell of the whisky Bill Bain opened his eyes. He swept the flask aside. 'An operator is forbidden to take alcohol when on duty,' he said, and collapsed again. That might well have been a last saying. But far from it – Bill Bain continued to feed the House of Commons annunciators for many years, as he had already done for so long. He set up in type the names of every Prime Minister from Asquith to Wilson.

The Company's aim was business as usual, and this was largely achieved – with of course a different emphasis on the various services. The records and a number of the staff were evacuated to Tring in Buckinghamshire. But a good many people remained in London, the Chairman and Managing Director among them. This division must surely have led to difficulties, but it is not reflected in the records. Nothing, for instance, could be more 'business as usual' than the Board-room minutes. There were discussions with the trade unions, a cost of living allowance, modest increases of salaries and wages – nothing individually worth a mention.

The working of the Pension Fund was naturally affected by the war. For those who were with the Forces it was difficult or impossible to keep up peace-time weekly or monthly contributions. Early in the war the Board decided to pay their contributions for them for the current quarter. At the end of this period the matter was brought up again, and the Board's decision was extended for another three months. So it went on. Not until December 1944 did the Board decide to continue their support until the men concerned returned to the office as civilians. At any rate, their contributions were paid throughout their military service. The ladies of the staff asked that they should be made eligible for membership of the pensions scheme. But the Company was not yet so far advanced in thinking to allow that.

The B.B.C. had completely and finally come into its own with the declaration of war, and made a new agreement with the agencies, paying £25,000 a year to be divided in the proportions: Reuters thirty-seven and a half per cent, P.A. twenty-six and a half per cent, Exchange Telegraph twenty-six per cent, Central News ten per cent. There was little cause of complaint here. It may be said at once that, whereas the first World War had been a slack and lean period for the agencies, the second was hard-working and not without reward. This was due to broadcasting and the realization that a full news service was essential to morale at home, and also a useful weapon abroad – call it propaganda if you will.

The Ministry of Information held conversations with General Anderson, mainly about news for the neutral countries. In the other services some domestic contracts were lost, but there were achievements abroad. Perhaps the most remarkable was that Osaka Mainici of Japan asked for the column printer service – shortly before Pearl Harbour. The Company started the war with an estimated rate of revenue of £280,000 and, after a big drop in the middle, ended it with one of over £300,000. But expenses were heavy. The losses of the first two years totalled £51,000, and later annual profits were much too modest to rectify this. In fact the Exchange Telegraph expended more than half of its reserves during the war. Salaried members of the staff were asked to accept a cut in their salaries.

Much the most serious casualty sustained in the 1914-18 war had been the death of Higgins. In the middle of the second war history repeated itself, although it must be admitted it had held its hand a long time to do so. On 11 February 1943, Lord Sandwich who was in the chair for the Board Meeting that day reported that Sir Wilfred was ill. The Board gave General Anderson full powers to act as Managing Director during his absence, but that was the only indication that the illness might be serious or prolonged. Wilfred King had been attending Board Meetings since he became Assistant-Secretary (in all but name Secretary, for no one held that post) in 1883. During those sixty years one could count on one's fingers the number of meetings he had missed – apart from those held during his summer holidays. There is no record of him ever being absent from illness before: rather it was because he was abroad or on a tour of the provinces.

But eleven days after that Meeting on 11 February, he died. At the next Board Meeting, on 25 February, with Lord Sandwich in the chair, the following was minuted:

*The late Chairman and Managing Director.* The Board wish to place on record their deep grief at the loss sustained by the Company in the death on Monday, 22 February 1943, at the age of eighty three years, [the death certificate gives his age as eighty-two] of Sir Wilfred King who had spent his working life with the Company, having served consecutively as Assistant Secretary (1883), Secretary (1884) Managing Director (1898) and Chairman (1913).

His direction and understanding of the various sides of the

Company's activities showed outstanding ability and his power to make friends and keep them and smooth out difficulties in competition were a feature of long and straightforward endeavour and of successful leadership of the Company to whose progress he had devoted himself.

He was possessed of boundless energy which opened up new avenues of activity for the Company and his vision carried the Company through adversity to the success of its manifold activities.

His human understanding and personal interest in the staff endeared him to his many friends who served with him and he will be sorely missed by them as he will be in his Office and the Board Room.

*Funeral Service* at Crematorium, Putney Vale, Thursday, 25 February.

*Memorial Service* at St Michaels, Cornhill, Thursday, 11 March.

After that tribute it was business as usual: The King is dead, long live the General. S. M. Anderson was at once elected Chairman and Managing Director. Charles Wills became a Director. He had already been on the staff for forty-six years. In 1919 he had become Assistant to the Managing Director. More than anybody else he deserves to be styled the friend of the man who at last had died. Certainly the new Chairman was grateful. Wills became his mentor, passing on those subtle yet important things which Wilfred King had compulsively kept to himself.

There is not much to be said about the war years after that. On 15 June 1944 General Anderson reported to the Board 'on our arrangements for coverage' of the Invasion of Europe. A limited pension scheme was in January 1945 approved for ladies who had served for twenty years.

But the great occasion was the Board Meeting on 10 May 1945. Then the Chairman and Managing Director made the formal announcement that hostilities had ceased in Europe at midnight on 8 May. Also that W. H. Kelland of the Copenhagen office had reappeared safe and sound. Edward Gilling, who was in the building, was brought in and congratulated on his Mention in Despatches. But there was also the news that W. E. West, the Company's correspondent with the Fourteenth Army, had been killed by a Japanese sniper. War news can never be entirely good, and hostilities with Japan still continued.

# CHAPTER 12

# House-Hunting

THE STAFF were given two days' holiday to celebrate the victory in Europe. Then they got down to work again. The special war services naturally came to an end, the Ministry of Information announcing its withdrawal of support of the successful Zurich office as from 30 September 1945. (It was in fact kept going for five years longer.) But news services were required until demobilization was completed. The War Office asked for a full bulletin which was to be supplied over military circuits to the British Army of the Rhine and elsewhere. The Political Intelligence Department of the Foreign Office called for general and parliamentary news for the Allied Press service. For a short time there was also a news service to Norway, Sweden and Denmark.

Some of the Company's men were honoured for their war service. General Anderson who had been Fire Guard Commander received the Freedom of the City of London. André Glarner was given the Croix de Guerre. Edward Gilling, who already had the M.V.O. for his pre-war work as Court Correspondent, was awarded the O.B.E. for his achievements as war correspondent in North Africa, Sicily, Italy, on D-Day, and in the subsequent Normandy campaign.

Others of the staff got into the minutes for the first and last time by dying or retiring after long service. Among the former were S. Smith who had worked in the Glasgow office for fifty-four years. Among the latter were operator J. Campbell and S. E. Mullins who were congratulated in the Boardroom and presented with £50 on completion of fifty years with the Company. But the record was put up by part-time Inspector Thomas Bradshaw who retired at eighty after sixty-six years' service, sixty of them on full time. Howard Bridgewater of the financial department also retired. Although the records tell little of him he appears from personal accounts to have been an erudite person who spent his spare time in proving that Bacon wrote Shakespeare. A silver tea set subscribed for by the staff was given to him when he left. On 13 March 1947 Charles Wills completed his half-century with the Company and was presented with a motor-car.

This was the year of the Exchange Telegraph's seventy-fifth anniversary. It was celebrated with an £11,000 bonus for the staff. The occasion marked not only the age of the small Company whose history we have been following but also its birth as a large Company. Although, of course, one cannot date this change precisely, it may be said to have occurred at about this time. The Exchange Telegraph was fairly well off, with £100,000 invested and £30,000 on deposit and an estimated rate of revenue of over £400,000. It had paid a fifteen per cent dividend the year before. It had recently acquired from the P.A. the whole of the Association's interest in the Central News, (formerly shared between them), and Column Printing Company. The Exchange Telegraph took over the Central News advertising, the P.A. its photographic library and Parliamentary service. But a clearer sign was that head office had grown too big for 62-64 Cannon Street. The sale of these premises was discussed, dependent on finding the site for a new building in a suitable area.

There followed a house-hunt which went on for years, punctuated by complications and disappointments, but not unprofitable in the end. It is necessary to tell the story as it were in serial form, or it would far outrun other activities.

The search, which was decided upon on 15 May 1947, seemed as good as over within a couple of months. Fawdry and Evans, the Company's surveyors, almost immediately found two sites in Cannon Street which were for sale together for £77,500. On 10 July the Managing Director reported that his offer of £64,000 for site 'A' and £12,500 for site 'B' – £76,500 in all – had been accepted and the solicitors hoped to arrange completion before the end of the month. There was also a site 'C' –

small enough for only £1,500 to be offered for it – about which the Managing Director had no information.

There then began a series of operations as complicated as some party game like Monopoly which the reader may care to follow as a mental exercise.

By the end of July, site 'B' had been bought and the contracts exchanged. The deposit had been paid for site 'A', and nothing had yet been heard about site 'C'. The acquisition of site 'A' was completed in September and it was decided to commission an architect to prepare preliminary plans for the new building – 2-6 Cannon Street. At the same time three offers were made to buy 62-64 Cannon Street. But none was satisfactory. It was resolved not to accept less than £225,000 for an immediate sale.

In October the Surveyors reported that they could get site 'C' for £1,500 and they were instructed to do so. They in fact bought it for less. By this time Messrs. Whinney, Son, and Austen Hall, the architects, had produced the first rough plans, which the Board studied. But the Managing Director then informed them that he had received a notice of compulsory acquisition of some of the land by the City Corporation. To this the Company lodged an objection – to no effect, for it was ruled out at a public enquiry at the Guildhall in January 1948. The land was lost, and the remaining land was unfit for economic development because of its awkward shape and limited extent.

The Board set about making this good. On the suggestion of the architect negotiations were begun for a strip of ground fifteen feet deep on the southern side of the proposed building site. This was designated site 'D'. The first difficulty was that the owner could not be traced. A greater difficulty, or at least alarm, was caused when the architect stated that the City Corporation might compulsorily purchase more of the site of 2-6 Cannon Street. It did so – 'for street-widening'. The Company put in a protest.

While this was still in abeyance it became known that the United Kingdom Provident Institution were prepared to sell their land, which adjoined the site, for £20,000. The Company snapped this up. In November 1949 the purchase of the additional land on the Cannon Street site known as 25 Old Change seemed as good as completed.

Now, in February 1950 two years from the beginning of the hunt, the architects were able to start on new plans – 'which can be subject to alteration as and when required at a later date,' the Board stipulated cautiously. But the land-hunger was still upon them, and for £4,500 they bought 61 Knightrider Street at the rear of the 2-6 Cannon Street site, and so created an island site much larger than the land originally purchased.

A year after that, in February 1951, the Company was still awaiting development permission. It was then leaked through Messrs Fawdry and Evans that the City Planning Officer was waiting for the Company to accept the compulsory purchase of the strip of land required for street-widening. They had made ample provision for this loss by purchasing other land to a total value of nearly £25,000. And they had still maintained their protest. They now withdrew it, and received payment from the City Corporation of the purchase price – 10s. (It was worth more like £30,000.) After that they got their development per-mission. But planning conditions continued to be imposed – for instance on height, to allow a person on Southwark Bridge to have an uninterrupted view of St. Paul's dome.

We must now go back to 1947 and see what events have been skipped over in our preoccupation with house-hunting. First of all, head office was not alone in its uncertainty about accom-modation. Throughout the war a monitoring station had been maintained at 16 Fitzjohn's Avenue, Hampstead. A year or so after peace returned this foreign language-speaking staff received notice to quit, and in 1947 moved to 15 Cannon Place. The name is a coincidence: Cannon Place is in Old Hampstead. The Panton Street office was sold for £66,000. (It had been bought for £11,000 in 1909), but – a condition which can only be explained by shortage of space – the Company continued to occupy the premises under the terms of a twenty-one-year lease. (The Exchange Telegraph in fact only rented it until 1956 when the Savoy Hotel took over the lease.) Nos. 36-38 Whitefriars Street was bought for £19,500 to house the engineers, and 85 Tooley Street, Bermondsey, was rented as a store. To save space at head office, the advertising department was transported to 17 Moorgate. There the Central News advertising agency joined it, the two having been merged in January 1947.

As horrific relief from these domestic details one may mention two manuscripts of some rarity value which the Central News brought with them in their files when they joined up with the Exchange Telegraph. These were a letter and a postcard from Jack the Ripper. Jack the Ripper operated (almost literally in both senses of the word) in a one-square-mile area of Whitechapel between 8 August and 8 November 1888. During these three months he murdered six prostitutes. All but one were killed in a street or alleyway where he might have been interrupted at any moment; and yet in all but one case (a different exception from the first) the body was mutilated in a particular manner and with surgical skill, one or more organs being removed – which must have taken some time. The reason why one victim was not mutilated was that the Ripper had to run, seconds after the murder. The exception from being killed out of doors was the last victim, pretty twenty-four-year-old Marie Kelly, who was found naked on her bed, her ears and nose cut off, most of the organs removed from the body and the heart placed on the pillow beside the almost severed head – an operation which, it was expertly stated, must have taken more than an hour.

The square mile of Whitechapel was saturated by the C.I.D. and policemen. Even thieves and pickpockets co-operated in the search. The more respectable citizens formed a vigilante committee. Various theories were advanced – and resulted in no more than the temporary arrest of innocent people. One was that the murderer used a long-bladed knife – so a half-witted Polish Jew nicknamed Leather Apron who used long-bladed knives for his work was picked up. But Jack the Ripper went on murdering and scientifically operating on his victims in the dark, and he was never caught.

For whatever reason, he wrote a letter and a postcard to the Central News, then a leading agency supplying the principal newspapers. The letter is addressed to 'The Boss, Central News Office, London City'. It is dated 25 September 1888. It is postmarked London E.C.3, two days later. It runs:

Dear Boss

I keep on hearing the police have caught me but they wont fix me just yet. I have laughed when they look so clever and talk about being on the right track. That joke about Leather Apron gave me real fits. I am down on whores and I shant quit ripping them till I do get buckled. Grand work the last job, I gave the lady no time to squeal. How can they catch me now. I love my work and want to start again. You will soon hear of me with my fancy little games. I saved some of the proper *red* stuff in a ginger beer bottle over the last job to write with but it went thick like glue and I cant use it. Red ink is fit enough I hope – *ha ha*. The next job I do I shall clip the ladys ears off and send to the police officers just for jolly wouldnt you. Keep this letter back till I do a bit more work then just give it out straight. My knifes so nice and sharp I want to get to work right away if I get a chance. Good luck.

Yours truly

Jack the Ripper

Don't mind me giving the trade name. Wasnt good enough to post this before I got all the red ink off my hands curse it.

No luck yet. They say I'm a doctor now – *ha ha.*

The postcard, written the day after he had killed two women, is almost illegible from bloodstains but just decipherable:

I was not codding, dear old Boss, when I gave you the tip. Youll hear about saucy Jackys work tomorrow. Double event this time. Number one squealed a bit, couldnt finish straight off. Had no time to get ears for police thanks for holding last letter back till I got to work again. Jack the Ripper.

To return to matters more directly concerned with the Company, fast-speed Creed teleprinters were installed to replace the existing instruments in the Stock Exchange subscribers' offices. E. A. Evans of the Company's engineering department invented a device by which the speed of the Creeds could be still further increased, but this was never put into manufacture. General Anderson lunched with the Council of the Stock Exchange and was later sent a copy of a resolution stating that 'the Council have no intention in view of terminating the Agreement with the Exchange Telegraph Company or the very friendly relations which have existed for so many years'. As proof that this was more than mere politeness, the Agreement of 1909 was not felt to need any revision until 1955.

Relations with the Press Association were as friendly as those with the Stock Exchange, both being justifiably pleased that the Joint Service was bringing in £50,000 a year to each. But a flutter was caused in the Company's Boardroom when it was learned from the draft of written evidence prepared for the Royal Commission on the Press that the P.A. were the holders, through nominees, of seven hundred and eighteen Exchange Telegraph 'A' shares. It was later learned that their broker had instructions to buy up any of these shares which became available. But no sinister motive was discovered: probably the P.A. just wanted a sound investment.

The Royal Commission on the Press sat throughout most of 1947 and 1948. First, questionnaires were sent to newspapers and agencies. Then proprietors, members of the boards, managers, editors and executives in general were interviewed. Innumerable questions were asked and answered. Each day's sitting was taken down verbatim and printed. The number of words published must far exceed those in the *Thousand and One Nights*. One needs courage to plunge into this ocean of verbiage, but it has been done, and the following extracts brought to the surface for their interest as concerning the Company's then aims and policy, also for the sidelights which are provided on certain characters. The questions were asked by any one of the distinguished gentlemen on the Commission. But for simplicity they are all here prefaced by 'Q'.

Q: Do you go beyond Europe and Asia? Have you people in the United States for example?

Brigadier-General Anderson: Yes, we have, in order to cover everything. In Washington, New York, San Francisco. In Europe we have men in Paris, Berlin, Rome, Brussels, Lisbon, Madrid.

Q: Practically in every country? – Yes.

Q: Do you go into Eastern Europe?

Wills: We have a man in Moscow.

Anderson: And in Prague and Athens.

Burn (the Editor): In Budapest, but not in the last few weeks.

Q: And in Asia?

Anderson: We have a strong representation in India, and two correspondents in China; and we are represented in Burma, Singapore and Baghdad.

Q: You have twenty in all in Asia?

Anderson: That probably is an over-estimate.

Q: What about Australia or New Zealand?

Anderson: There we take a service from an agency, but we have one or two people from whom we can get information if we want to.

Q: What about Africa?

Anderson: Yes, we have a pretty complete cover.

Wills: And South America too . . .

Q: How do you recruit your journalistic staff?

Anderson: We look for them. My editor, if we want new reporters or sub-editors, gets among that type of person and we bring them in . . .

Q: What is your system of training?

Burn: A boy begins as a messenger in the editorial department. His job is to wait upon technical staff. Then he has two opportunities, to go out with reporters, to telephone their copy and see how they work, or to act as a telephonist inside the office to see how copy comes into the office and is handled there . . . From the position of internal or external telephonist he graduates at the appropriate age to junior reporter or sub-editor . . .

Q: Have you many reporters on your staff who have served on the staff of newspapers?

Anderson: Reporters and sub-editors, yes: a considerable number.

Q: Is it a different kind of job?

Anderson: Yes, it is more factual. Another characteristic is that there is no hour at which one goes to press. Exchange Telegraph goes to press every minute of the day and there is no leading up to a peak hour and then dropping off: the training is to that extent quite different. Another characteristic of our services is that we have no political or other orientation; it is purely factual . . .

It may be difficult to believe, Sir Geoffrey [Vickers], but I have been with the Exchange Telegraph a long time and I have no political conscience at all. My reaction to public pronouncements is entirely neutral. That is a frame of mind which I endeavour to inculcate into the whole of the staff . . .

Q: I gather you bought Central News Limited in 1937 and for some time ran it at a loss?

Anderson: We bought it in conjunction with the Press Association.

Q: And with the Press Association ran it really in competition with your own service and at a loss. What was the reason for that?

Anderson: It had been a well-known agency and had done quite well for a long time. Then it began to fail and it did not get enough support, it was on really bad times. It was largely owned by a Canadian, and he asked us if we would like to buy it, the point being that there was a foreign agency in the market to buy it, and we objected to that, so we bought it. We thought it was the right thing to do. We tried hard to resuscitate it but we could not, and then times got worse and the war came, and it had eventually to stop its foreign service, which was too expensive, and then it had to cut down and stop its home service. That left it only a Parliamentary service and an advertising service . . .

Q: Do most of your staff stay with you, or do they flow from you to the service of the newspapers and back to you with the same tempo as journalists generally pass from one newspaper to another?

Wills: No, they do not.

Burn: Long service is characteristic of the office.

Q: Few journalists who have given evidence here have been with one paper all their lives?

Wills: We had a picture taken last week of twelve people who have each done more than forty-two years or forty-three years. One of them has done sixty-one years and he will not go . . .

Q: The company is a public company?

Stevens: It is.

Q: Are the shares dealt with on the Stock Exchange?

Stevens: There is a very limited market on the Stock Exchange.

Q: That is strange, is it not, in view of the profits of the Company?

Anderson: The number is small.

Stevens: There are just over one hundred shareholders and most of the present shareholders are the descendants of original shareholders or have come in with the break-up of original bigger holdings; they are mainly family holdings. They are mostly small holdings and they are not shares that one would normally speculate in. There has been no speculation in this company's shares in my experience.

Q: It would be possible, theoretically, for the shares to be acquired by a group?

Stevens: Anything is possible. It would be an extremely difficult task.

Q: There are no transfer restrictions? – None.

Q: I think it is correct to say, is it not, that the major portion of the income of the Company is from private subscribers?

Stevens: I would say that somewhere between seventy per cent and seventy-five per cent of the total income is from subscribers other than newspapers.

Q: The revenue from private subscribers appears to be three quarters of the total – Yes.

Q: So your main business is with private subscribers, is it not, and not with the Press?

Stevens: Our main business is the collection of news primarily for the Press. As a subsidiary activity, and in order to cover the cost of the news services, we sell this news in as many markets as possible. That brings us into the happily fruitful market of selling news to private subscribers, very largely sporting and financial, from whom we collect a considerable revenue, which enables us to spend more money on perfecting the news services. The subsidiary activity has become bigger in money value than the primary activity,

P

but nevertheless the primary activity continues to be the collection of news for the purposes of the Press.

Chairman: Thank you very much, gentlemen.

As to the findings, only a hundred words or so can be quoted here. After listing 'the principal news agencies' – P.A., Reuters, Exchange Telegraph, British United Press and Associated Press – the Commission stated:

> There are other small specialised agencies, but the 'big five' are the main sources of news – both home and foreign – for the principal newspapers of the country (dailies and evenings) and also for the B.B.C.
>
> The Associated Press and the British United Press are extensions of the two principal American news agencies.
>
> The other agencies are British controlled, but are now in large measure financially integrated concerns *from which the elements of competition have been removed.*

The words here italicised have in the Company's file copy been underlined with a thick red pencil and are dismissed by a large exclamation mark. One feels that no one below the rank of Brigadier-General would dare to do that to a Royal Commission's report.

On 13 November 1947 the Board decided to cease publication of the Economic Bulletin 'due to lack of support in the present restrained market conditions'. The chief interest in this minute is that it is headed, 'Extel Economics Bulletin'. This is the first occasion that the present form of the Company's name appears in the records. It is an isolated example, for 'E.T.C.' was used for some time thereafter.

Several of the Directors were given indefinite leave due to ill health at this period. Each returned to duty after a month or two. But on 6 April 1949 Stanley Christopherson died. He was eighty-seven and had been on the Board for thirty-five years.

> In their tribute to him his colleagues stated: Although possessing wide business and other interests Mr Christopherson always gave to the Company a full measure of his time and wide experience, and his wise counsel will be greatly missed.

It was more than a formal minute. In every aspect of his work, from representing the Company on the Joint Committee to backing up or restraining the enthusiastic Charles Wills in the affairs of the Sports Club, Stanley Christopherson had proved his absolute reliability. He had lived a full life. He had been Chairman of the Midland Bank and High Sheriff of Norfolk. He had been on the Board of a large number of companies. He had played cricket for England against Australia and was President of the M.C.C. 1939-41.

Only a few weeks later Wills retired as Manager. He remained on the Board and in the post of Chairman of the Column Printing Company which he had recently assumed. But this meant the end of his daily attendance at the office which he had maintained for fifty-two years, twenty-one of them as Manager.

From 1 June 1949 William C. Stevens was promoted to General Manager, and T. F. Watson became Secretary, with E. H. Martin as his Deputy. This period – under Stevens – was spent in close examination of revenue, and the cutting out of unprofitable services and the improving of others with consequent increases in subscription rates.

Thomas Watson has only previously been mentioned as representing the Auditors, but like W. C. Stevens he had joined the Company from Deloitte's, having already become thoroughly acquainted with the financial side of its business from close association with it during many years. Watson had, on joining, become the Company's chief accountant with G. Wittleton as his assistant. Soon afterwards Leslie Cross joined the Accounts Staff. Watson had acted as Secretary during Stevens's absence through illness, and in June 1947 had been appointed Secretary of the Pension Fund. As such he piloted a new Pension Fund Trust Deed.

In October 1948 the Company had taken over Thames Paper Supplies and turned it into a Limited Company. There will be more to be said about this in another context. But it is appropriate to mention here that Stevens became Chairman and one of the three Directors of the new company, and Watson the Secretary. Rather more than two years later, on the death of Robert Parker who had formerly been the Proprietor of Thames

Paper Supplies and latterly a Director, Watson was elected to the Board, and D. W. Drakeley, who is still with the Company, became Secretary. The latter is now Managing Director of Thames Paper Supplies.

A paragraph about the wage-earning staff: since the seventy-fifth anniversary bonus their earnings had gradually increased, partly of course by personally won increments but also through trade union bargaining with the management. But one incident will show that the old spirit remained. H. Southern, a Law Courts reporter, asked permission to retire on pension. His request was considered by the Board on 10 May 1951. 'It was agreed that in view of the fact that he is now seventy-eight years of age this request should be granted.'

We have reached the point where the story of the Company's adventures in real estate was interrupted. It will be remembered that development permission had at last been obtained, and Austen Hall was preparing plans of the new building.

On 10 May 1951 General Anderson read to the Board a letter from the surveyors. It stated that 'certain complications had arisen in connection with the development permission which had been granted by the Corporation of London'. What these were is not recorded, but a month later the architect was told to 'proceed with the preparation of detailed drawings'. This was in spite of the fact that the final plot of land required to complete the building site had not yet been obtained. By September, however, this had been done. The 1,869 square feet of bombed site known as 21-24 Old Change had been bought for £5,230, and this included the licence of the *Crown and Sceptre* public house which had formerly stood there.

The detailed plans were completed and circulated. The Board liked them. But the Town Planning Officer objected to the continuous balcony on the first floor. The architect was instructed to protest strongly against any reduction in the length of the balcony. A note of impatience with bureaucracy is at last evident.

In March 1952 the Exchange Telegraph celebrated its eightieth birthday. General Anderson was at this time conferring with a sculptor about the design for the front of the building. In September he reported to the Board that the latest design for the exterior had been approved by the City Planning Officer and that there was every prospect that the Town and Country Planning Authorities would accept it.

In this same month André Glarner retired from his post of Manager of the Paris office and was replaced by M. Lusinchi from Geneva.

In December 1952 application was made to the Licensing Justices that the licence of the *Crown and Sceptre* which formerly stood on 22 Old Change should be granted to William Charles Stevens who would hold it as nominee of the Company.

There followed a period of more than a year without any record of activity concerning the building. It must be most unsatisfactory to be the licencee of a pub that is not there. Best to turn one's attention to other matters.

The Pension Fund ventured into the estate business by buying a block of thirty-four flats in Canons Park Close, Edgware, helped by a loan from the Company. The Fund was also, whenever possible, buying Exchange Telegraph shares. Nearly all of these have since been sold at a handsome profit. The Company had recently increased its capital to £250,000.

In June 1953 the service of the actuary, Owen Kentish, was terminated. He had worked for the Fund since its inception twenty-four years before. In his place the firm of R. Watson and Sons was appointed.

Later in the year the Newcomen Street mechanics moved to their new workshop at 13-19 Curtain Road. This was also the home of Thames Paper Supplies, Ltd, which increased its usefulness by buying a slitting machine, thus becoming able to make tape for the Company's instruments. The Manchester office moved to new premises at 67 Market Street.

The foreign news service was shrinking, its directly post-war purpose being ended. But R. W. V. Parker was sent to Bermuda to report the Three Power Conference of 1953, and Louis Nickolls, who had become Court Correspondent in 1945, sailed with the Queen and Prince Philip in the *Gothic* on their round-the-world commonwealth tour. In Paris, André Glarner died at seventy-one, not long after his retirement.

The Directors, having no trade union to look after them, voted an increase in their fees. And their number was increased by the appointment of John Glyn, a Director of Glyn, Mills, the Company's bankers. But a bigger change was foreshadowed when General Anderson announced his intention to retire from the position of Managing Director on his seventy-fifth birthday, 23 September 1954. He was persuaded to retain the Chairmanship. In any case the family representation was assured, for the General's son, D. G. Anderson, was on the staff, at this time as Assistant Services Manager in London. Captain Douglas Anderson had had a varied war career, largely in Italy where, by an interesting coincidence, he was for a time instructed in mountain warfare by the author of this book. He was one of the very few officers who gained a distinction, and also one of the somewhat larger number who won an Italian wife.

The new building for 2-6 Cannon Street was once more under discussion in February 1954 when another set of drawings was produced by the patient Austen Hall. But he received a severe shock a month later when his revised estimate of costs was considered. The Board stated that in view of the great increase in these costs it was extremely unlikely that the Company would be able to finance the venture. Consideration was given to selling the site and making do with a slightly altered old head office.

The end was not yet, but we may as well anticipate it. In March 1955 the site of 2-6 Cannon Street was sold for about £¼ million. Thus the work of putting together a jigsaw of plots of land which had begun eight years before came to an end, not as originally intended but with a very handsome profit. Edgar D. Evans of Fawdry & Evans had negotiated the sale as efficiently as he had negotiated the purchase of the various plots which eventually became the site.

# CHAPTER 13

# Extel House

GENERAL Anderson had said that he wanted to retire on his seventy-fifth birthday, in the autumn of 1954. Instead he died in harness on 23 May. He had faced a difficult task in taking over from Wilfred King, but during his eleven years as Chairman and Managing Director the Company had done well. Also, he was probably the most personally liked of the men who had headed the Exchange Telegraph. His business friends, his colleagues on the Board, the staff – everyone had respected and been fond of the old soldier. He was buried near his beautifully placed home at Studland, beside the sea.

William Charles Stevens was elected Chairman and Managing Director – in time to take the final decision *not* to build on the 2-6 Cannon Street site but to sell it instead. He deserves the most credit for these financially successful negotiations. T. F. Watson was appointed a Director with the title of Assistant Managing Director. Eric Martin became Secretary.

One of the first decisions under Stevens's Chairmanship was to buy Victoria Blower Co. (1942) Ltd. One of a number of shareholders, Alfred Cope, was offered a place on the new Board, the other Directors being W. C. Stevens (Chairman), T. F. Watson and G. H. Wray, the Exchange Telegraph's newly appointed Services Manager.

The Victoria Blower has a most intriguing name – using the adjective in the journalistic sense. When, some years later, this company became sufficiently famous to have a horse race called after it, a lady wrote to the *Daily Mirror*:

The Wolverhampton races were on last month and I noticed a race called the Victoria Blower Handicap. As my

mother's name was Blower and she came from Wolverhampton I'm interested to know whom this race is named
after.

Dickens might well have written about a Mrs Victoria
Blower. But the name comes from the vernacular word for a
telephone. How did the telephone get this name? Because the
original telephone was a speaking tube into which one had to
blow – causing a whistling sound to call attention. The Victoria
Blower is a telephone service for bookmakers. That is the short
explanation. To attempt anything fuller would involve mathematical complications which could only be understood by a
senior wrangler, or a racing man who knows the answer
already. But to give a hint to the uninitiated, one paragraph
may be quoted from an article by Jim Adams, past Manager of
the Victoria Blower Co., in the Extel News:

> There are occasions in the life of any office bookmaker
> when he finds himself holding too much money for one
> particular horse . . . There are one or two courses of action
> open to him. He can either (a) shift it with or without com-
> mission to another bookmaker at S.P. (Starting Price) or
> (b) send this excess money to the racecourse to be invested
> at the current price, in the hope that this will prove to be
> greater than the S.P., giving him a margin of profit on the
> actual bet. Thus a bookmaker sending £50 to the course
> and obtaining 4-1 about an eventual 3-1 chance receives
> £200 from the course out of which he has to pay his
> punter(s) £150.

So much for explanation. But there has been a good deal
about racing in these pages, and we have not yet been *on* a
course. Before returning to the reconstituted Board of Directors
there is time for a lively glimpse of a Victoria Blower team at
work. It has been lifted from the files of the House Journal and
comes from the pen of its Editor, Louis Nickolls. The
characters are *not* imaginary.

The scene is Sandown Park racecourse, fifteen miles out
of London, on the Portsmouth Road. It's a cold, gloomy
March day with rain threatening as the time for the first

race approaches and the Victoria Blower boys take up action stations.

In the front of the grandstand, Charlie Wilson clamps his telephone headset on more firmly, contriving to adjust the top piece neatly over his tweed cap.

'Hello, Betty!' he says, over the open line.

In a corner of the Sports Room in London, Mrs Mary Betty is at the other end of the line. She passes Charlie a code message from a Victoria Blower client: *Vic. 500 Retz. Signed Joe.* 'It is the last two words that are significant here,' explains Charlie. 'It means something – but we don't know what,' he says.

Standing close by Charlie's shoulder is H. Alexander. Everybody calls him Alec. Nobody seems to know what the H. stands for. Alec's arms and hands move swiftly as he makes his tic-tac signals down to Tattersall's ring. He sends the message in slightly different form, using the horse's number as shown on the racecard, instead of its name, Retz.

Down below the stand, in Tatts, the Blower 'floor men' are waiting, standing round in a small, expectant covey with Johnnie Gothard at their head. It's Johnnie who reads the tic-tac signals. Then he, too, breaks into the tic-tac rhythm, repeating the message to check that he has got it correct. Having got the O.K. from Alec, he passes the message on to Bert Gandar who is standing beside him. Bert fades into the race crowd, to return a few minutes later. Now Johnnie Gothard is signalling again, back to the stand: *Vic. 1250 to 500.* That's the bet; a 'monkey' (£500) at 5-2.

Back goes this information over the open line via the check-capped Charlie to Mrs Betty.

Other messages are coming back by high-speed tic-tac from the ring. Charlie Wilson keeps on talking over the 'phone to the listening-post at Extel House:

*So far, he's got 550 invested, Betty.*

*800 invested so far, Betty. 250 to come. Right?*

There's a shrill, ear-piercing whistle from Johnnie Gothard to attract attention. His arms and hands begin to talk again.

*O.K. Betty. All on. 1050.*

It's all go, as they say . . .
*They're running, Betty. Nothing done.*
*No more. That's his lot.*
*Can't lay any more, Betty.*
*He's got 500 at evens.*

Here and there, the bookmakers' codes add an odd note to messages. One of them comes onto the 'phone to Head Office and says: 'It's fish this week'. So, a little later, the mysterious message *Hake 105* is transmitted by word of mouth and tic-tac, to and fro. Another week, this same bookmaker uses a 'fruit' code. Others use colours, the names of birds and trees.

Here we go again: *Binns to his people. Sid ring Ted.* Off goes another of the floor men.

In a little while, back comes the message from the ring: *O.K. Binns delivered.*

*Flag up, Betty.*
*They're running.*

A man on the rails in Tatts is still signalling furiously. 'That's Mickey Fingers,' says Charlie. 'He's an agent. Offers prices from the rail bookmakers to the board bookmakers. Nothing to do with us.'

While the race is being run, particulars of twelve separate bets that go to make up that total of £1,050 are being passed from the ring. By tic-tac, of course. Charlie puts them over the 'phone:

*200 at 9-4 William Hill, Betty.*
*50 at 9-4 Izzie Isaacs.*
*100 at 9-4 Victor Chandler.*
*50 at 9-4 Bob Stock.*
*100 at 2-1 George Dexter.*
*50 at 2-1 Bernard Howard.*
*50 at 2-1 Duggie Wilson.*
*150 at 2-1 William Hill.*
*50 at 2-1 Victor Chandler.*
*100 at 2-1 with Jackie Cowan.*
*100 at 7-4 Briggs Berman.*
*50 at 7-4 Alf Turner.*
*1,050, Betty. O.K. Betty?*

'With the names of clients, we sometimes use a single sign instead of spelling them out by tic-tac,' says Alec. 'Did you notice *Binns,* for instance. I put both hands up to my eyes, as though I was looking through racing glasses. Binoculars. Binns, for short. When Jack Burns was alive, we used to pass his name by sucking a finger and wringing it as though in pain from having burnt it. Then we used to have a client named Sutcliffe. We had a forward cricket stroke for him. It all saves time. William Hill; we make a sort of up-hill-and-down-dale sign for him. For Day, we open our hands and look up at the sky as though to say: "What a wonderful day!"'

But this is no sort of day at all, at all, as they say in Ireland. The drizzle suddenly turns to a downpour. Up go the bookmakers' big umbrellas. On goes the floor men's wet weather kit – the mackintosh leggings and plastic macs.

'Rain is bad for the floor men,' says Charlie. 'You get a lot of people crowding under the umbrellas which makes it more difficult for 'em to work. Another trouble is trying to keep your card dry otherwise it messes up the writing. Even with walkie-talkie, you've still got to write and make the entry on your card.'

'Besides,' he says, 'it's pretty miserable out there when it's pouring with rain like this'.

Johnnie Gothard comes up to the stand, to a point just below where Charlie is standing. He tosses up a weighted piece of paper which Charlie, after making a good one-handed catch, unfolds and studies.

*13-8 the winner, Betty.*

'That's the S.P.,' he says. 'The starting price. He uses a penny to weight the paper. Now, he wants it back again. We always have to return the penny. All this'll disappear one day with walkie-talkie all over the place. Sorry we're not doing it by pigeon post. We've lost the pigeons.'

*No, love. Soon as it goes in the frame, you'll get it, Betty.*

'She wanted the fourth,' explains Charlie.

A good many people leave before the last race, the weather is so wretched. But the Victoria Blower boys never let up. They are still out there in the rain in Tatts, going round the

bookmakers' stands with their assortment of commissions. Alec and Johnnie Gothard continue to fling their mysterious, meticulous signals at one another across the void. And Charlie Wilson is always on the line:
*Okay, Betty.*

A major concern of the Board was still accommodation. They thought of building on to 62-64 Cannon Street, but gave this up on hearing that the Corporation of London planned to widen the street. They then looked for new premises, or a site to build on. Meanwhile they bought 36-37 Queen Street into which the Statistics' production and editorial departments moved.

No more than brief mentions have been made of the statistics service since the description of its successful start in 1919. It had continued to prosper and increase year by year, until the lean decade of the 1930s, when it began to fall off. The only item which remained steady was the printing cost. As stated when the statistics department was first described, a new card was issued whenever the company in question made any important announcement, or when one was made about it. The original records are all-important. The reader will therefore appreciate the chaos which was caused when Budge Row, which then housed the department, was burnt out during the war by a shower of incendiary bombs.

An estimated £2,550 worth of damage was done to the building and equipment. But that was of no great importance. You can build a house again, you can buy new printing machines. Such things in any case are insured. But the records were a different matter. A large number of valuable chestnuts had been picked out of the fire, but 4,300 items of copy were lost. What were they worth? The information still existed since the companies concerned still existed. But the data had to be re-collected. The Board of Trade Assessors for War Damage calculated the loss at £1 for each item, or 'kind of Copy', and a payment on account of £10,000 was a great help to the Company which was then in financial straits. During the two years which followed, the cards for which this copy had been destroyed came up for re-issue. The operation had to be carried out again from the beginning.

After this disaster the department was evacuated to Tring, where a skeleton apparatus and staff quarters had been prepared as part of the pre-war emergency measures.

The link with the United States was broken when Horace Leslie Hotchkiss junior died at the age of eighty-five in October 1954. He had been a Director for twenty-six years. With his father's still longer service added, this meant that there had been a Horace Leslie Hotchkiss on the Board almost throughout the Company's existence. It was decided not to appoint another American Director.

There were some changes on the staff during the following year. Eric Steadman, the chief engineer, had been ill for some time. During his absence Gordon Dain was appointed deputy chief engineer. And Edward Gilling, last mentioned as a war correspondent, moved up from News Editor to Deputy Editor. E. W. H. Bond became Assistant Secretary, Douglas Drakeley an Administrative Assistant, which means someone who can be given awkward problems by the Managing Director. And – not a change but an honour – the Court Correspondent, Louis Nickolls, was awarded the M.V.O. in the Birthday Honours. On the Board, T. F. Watson was appointed Deputy Chairman to take the place of W. C. Stevens during his visit to Australia and New Zealand to attend the Commonwealth Press Conference during the northern hemisphere winter of 1955–6.

So the Chairman was away when a possible new home was found by Fawdry and Evans. It was only in course of construction and as such cannot have looked very inviting, but it is evident that the Deputy Chairman and the other Directors fell in love with it at first sight. There is no record of them even discussing it. They told their agent to go ahead, 'subject to the settlement of certain outstanding points', admittedly, but to go ahead on the basis of a forty-two-year lease.

It was called Island House because it was on an island site, which suggests the South Seas, but means that it was cut off by streets from all other buildings. Thus it was comparatively isolated and certainly individual. It was a hundred yards north of Fleet Street, half way between Fetter Lane and Shoe Lane, and close to Dr Johnson's house. The only thing the Directors did not like about it was its name. They re-named it Extel House, even before it was a house.

Thus for the second time we find the word Extel. A good deal of research and questioning has failed to establish who coined it. Exchange Telegraph Company is obviously too much of a mouthful, and in fact the full name was very rarely used in all the Company's written history, except in such pompous things as prospectuses. Even in legal documents it never got beyond the first mention: it was 'hereinafter called the Company.' 'The Company' it was also called in internal communications. 'The Exchange' was the most common early abbreviation. It is said that 'E.T.C.', which was used for so long, was finally objected to by certain of the Directors because they did not want the Company to be thought of as a mere *et cetera*. Also it was liable to be confused with the Eastern Telegraph Company. It is said by several members of the staff that Gordon Dain, who became chief engineer when Eric Steadman finally retired, started using 'Extel' as a conveniently brief telegraphic word – which it is. But Dain himself disclaims this, and in fact the earlier use in connection with the *Economic Bulletin* puts it out of court. In the evidence collected by the Royal Commission on the Press the Company is referred to, in brackets after Exchange Telegraph, as Ex Tel – which suggests gradual evolution. The editorial department used 'Extel' before anyone else in the Company. In any case Extel House is very much there to see today, five storeys and a penthouse.

It was not yet there, however, at the stage of the story we have reached. And although the building was rapid and the occupation quicker there was other activity before it was completed.

It was suggested that the Press Association might join with Extel (at last we can use the word without fear of anachronism!) in the control of Thames Paper Supplies Limited. But Edward Davies, the P.A.'s General Manager, returned the answer that his Board considered it 'rather outside the normal activities of the Press Association'. They would, however, consider buying the products of Thames Paper Supplies 'if prices and quality were satisfactory.' This, in the long view, may be considered another happy negative for Extel. Both the P.A. and Reuters have bought Thames paper ever since.

In February 1956 N. Moschopoulos, the Athens correspondent, gave notice that he wished to retire because he was eighty-five. A new appointment was made, but it was of very short duration, for it was soon afterwards decided to close down the whole foreign service at the end of December 1956. (It had been losing over £20,000 a year.) This, though briefly told, was one of the dramatic changes of direction in the Company's aims which, by coincidence, was brought about soon after the finding of Extel House.

Individuals may seem of lesser importance in the context of big policy decisions. But the fate of every member of the staff is recorded in the minutes, so as many as possible should be mentioned here. In 1957 Frederick Wolstenholme died at eighty-two. Although he did not rise higher than assistant engineer he was one of the striking Extel characters. His daughter also served, and at this time had recently retired. Before ladies were made eligible for membership in the Pension Fund, she protested strongly against this state of affairs. The Company was poorer without the Wolstenholmes. Other casualties from age were Arthur Neat who had been chief operator of the financial service, Edwin Fallas, formerly head printer at Panton Street, who had joined the Company in 1889, and Howard Bridgewater, late financial editor.

In a lighter vein – although the stakes were high – we may turn to the Company's unusual flutter on the Grand National in the Spring of 1957. The story begins at least two years before. Since the ban on telephones on racecourses had been withdrawn it had been customary to pay as little as two or three pounds for permission to give a running commentary of a meeting and for using the telephone facilities. This was not niggardly; rather it was a nominal fee, for the racecourse gained from the publicity and gave away nothing material. In 1955 the Racecourse Association asked for very much higher fees from the Joint Service, suggesting that the Company and the P.A. should recoup their expenditure by acting as agents for the Racecourse Association for the collection of monies from off-course bookmakers. The Company and the P.A. refused to pay the extra fee or to co-operate in the plan, and the Racecourse Association withdrew its demand. The Company then offered £7 10s 0d for each racing day covered, with a minimum of £20 and a maximum of £150 in any one year; and this was the basis of the agreement arrived at. The arrangement is very different nowadays when the Racecourse Association annually receives from betting shop proprietors over £700,000 through the Extel-P.A. Joint Service commentary fund.

Today, it may be arguable that the Grand National is the most important race in the calendar, but it is surely the most exciting to listen-in to or otherwise follow from afar for the millions who cannot get to Aintree. This course, as is well known, belongs to Mrs M. D. Topham, who was not then a member of the Racecourse Association.

In February 1957 Mrs Topham and the Secretary of the Manchester racecourse demanded £500 a day for telephone facilities. They were informed that the Board declined to pay this sum, but that they were agreeable that no running commentary should be given on the Grand National that year. They offered to pay £250 per annum for the telephone facilities. The P.A. followed the same line.

Negotiations ensued and were unsuccessful. T. F. Watson went to tea with Mrs Topham. She was hospitality itself, but she was also adamant. So was the Secretary of the Manchester racecourse.

On 21 March the Board were told that Mrs Topham had further demanded that if telephoning was permitted (at her price, of course) only newspaper subscribers might be served, *not* private subscribers. The same conditions applied to Manchester. The Company and the P.A. refused both the price and the restriction on service. So did the London and Provincial Sporting News Agency. Consequently no telephonic facilities would be forthcoming at either racecourse. All subscribers were informed of the difficulties which had arisen and were told that 'every endeavour would be made to supply a satisfactory service by other means'.

It looked like a return to Alf Pepper's telescope, but Aintree does not lend itself to such methods. It is a large course with interest at every one of the thirty jumps, so a number of reporters are needed.

On the days concerned nothing unusual was to be noticed. There was no wire-tapping or surreptitious laying of telephone lines on the course. But an observant person might have been surprised that one of the Company's staff appeared to suffer so much from the cold. Although it was pleasant spring weather he was heavily wrapped up, and he bulged where the male body generally does not.

For some weeks past the Managing Director had been experimenting with the possibility of using walkie-talkie apparatus, and various tests had been carried out in London where it was thought that conditions would be far less favourable than on any racecourse or at any other sporting venues. George Wray, Sam Pepper and other members of the racing staff were involved in these tests, which included two-way speech from the roof of Extel House to the balcony outside the dome of St Paul's Cathedral. The tests were on the whole considered sufficiently satisfactory to justify using the apparatus on a race course and it was decided to proceed on that basis.

On 16 April the Managing Director reported to the Board that a satisfactory service of racing information was supplied to newspaper and private subscribers from the race meetings at Manchester and Aintree. This included the Grand National.

Let Freddy Haines, the man who bulged and seemed to suffer from the cold, fill the gap in this story in his own words:

> I was brought into this conflict when the Managing Director, Mr Stevens, called me into the Boardroom. I was asked to put on a harness which held a walkie-talkie, the size of a large biscuit tin, this enabled the walkie-talkie to be carried in front of the stomach. A raglan raincoat was produced and I put it on. The mike was passed down my right sleeve and the ear piece down the left.
>
> Manchester became the scene of our first conflict with the racecourse ruling. The late Sam Pepper had arranged to rent a front bedroom in a house overlooking the course. He had binoculars and walkie-talkie receiver and a colleague

Q

on the stairs ready to pass messages to the phone-man in the hall.

On the course I made my way to the rails in order to keep a clear route to Sam. The course boys, some of whom I had never met, fed me with a constant stream of shows [state of the betting] which were passed back to our bedroom office. For someone who knew very little of the racing news service, I felt I had been transferred to a new world.

It was after racing that I learned that all our efforts had fallen on deaf ears, for Sam had been able to pick up one of the course boys with his binoculars and had been getting a first class service from reading Micky Finger's shows. This was borne out later when we were informed that on that day the service carried more shows than normal.

Shortly after the Manchester meeting came the Grand National meeting at Aintree.

Once again Sam had worked his charm and had acquired a desirable residence with a garage into which we were able to secrete the car which carried the receiver. With the co-operation of the Post Office and members of the Liverpool staff, wires were run to the first floor little room – not ideal surroundings – but it had the required facilities, a clear view of the Grandstand and the Grand National start.

Prior to National day things had gone very smoothly despite Aintree's security staff who hounded the course boys throughout the afternoon and even tried to stop them using the public telephones.

Grand National day and the real test. The security men were really on their toes right from the second I walked through the turnstile, and carrying all before, this was not a simple matter.

The *Daily Express* had drawn a supposed facsimile of myself, with bowler hat (carrying batteries) and umbrella (the aerial), a real city gent who would have stood out like a sore thumb. The greatest disservice came from another morning paper which gave the car registration number, but by a stroke of luck we had managed to get the car into the garage and wired up.

Once through the turnstile I felt as if a thousand eyes were upon me and could see through the raincoat, and the walk to the front of the grandstand seemed a mile.

Only once during the meeting was there a moment of panic and that came early in the afternoon when one of the course boys called for me to 'move fast' and pushed me to the left, and to my surprise a tunnel through the bookmakers appeared and as I went through it closed behind me thereby stopping anyone who might be following.

After this incident there was a notable easing of the tension and the withdrawal of the security staff.

I must admit that on leaving Aintree after the last race, knowing that the service had not been interrupted, a great weight had been taken off my . . . stomach.

When the racing services report was given by the Managing Director, the Board were meeting for the first time in Extel House. It was eighteen months since they had found it, and it is interesting to follow the stages of its building and occupation. Before the agreement was signed, a panel heating system had been found and inspected at the Bankers' clearing house and ordered for the building. The Assistant Managing Director and the Secretary were authorized to sign an agreement and lease if necessary without waiting to report at the next Board Meeting. Here are signs of impatience which anyone who has waited long for a suitable house knows well.

The sale of 62-64 Cannon Street was as good as achieved. The date of the Joint Service Jubilee was reached. The Board recorded their 'feelings of sincere friendship and satisfaction' and celebrated as guests of the P.A. at a lunch at the Savoy on 11 October 1956. William Stevens was on this occasion warmly thanked for the excellent work he had done for the Joint Service.

But the Board's thoughts remained with Extel House as frequent references show. Conferences with the Post Office on telephonic and telegraphic communications were begun. It was decided to furnish the Boardroom 'on a suitable and dignified basis'. Running through the records, one begins to look out for the next sign of impatience which almost everybody shows – to start living in the place even if unfinished and interrupt the workmen at their tea.

Sure enough, commando groups of the editorial and sports services moved in on Saturday, 15 December, occupied the first and second floors, and set up communications so that the rest of their departments could follow close on their heels. The operation was entirely successful.

Nothing succeeds like success. Everything was going well. Both Reuters and the B.B.C. agreed without argument to pay substantially more for the services provided by Extel. The three subsidiaries, Central News, Column Printing Company and Thames Paper Supplies, were showing profits. Even the Victoria Blower was doing better than it had at the beginning, thanks to co-operative action by the Company and a committee of subscribers. Fast-speed Trans Lux was installed at the Stock Exchange. A Communication system using a newly developed tape machine was installed at London Airport. The capital of the Company was increased to £300,000.

During the weekend of 19-20 January 1957, the Administrative Departments moved into Extel House. It will be remembered that the building had been taken on lease. This, of course, meant paying a substantial rent. To avoid this inconvenience the Company bought the City Island Property Company, the owners of the site and building – a most satisfactory way to treat a landlord. And they arranged a lease of Pemberton House next door.

Having taken over Extel House completely, a start was made on providing amenities. A staff restaurant was opened in May and a managers' dining-room (for the heads of departments) the next month. There was more to be done, but at least the staff were being fed. When, in September, the engineering department was brought in from Whitefriars Street to Pemberton House across the way, the new headquarters may be said to have been established.

# CHAPTER 14

# Birth of the Group

IN JUNE 1957 the Board decided to purchase the sole manufacturing rights of a small portable radio telephone – a walkie-talkie in all but name; but such things were still rare outside military circles in those days. It was the invention of a man named McCabe of Nottingham. This was one of the occasions when the Board pounced like hungry tigers. In no time a company styled Extel Communications Ltd had been formed, with a capital of £1,000. The set was demonstrated to the Police and the Fire Brigade, and everybody was impressed – though nobody placed an order.

The new company did not achieve anything in the commercial field; it could scarcely have hoped to on its modest capital with giant companies all around with transistors up their sleeves. But it did several good turns to Extel. Although it appeared too late to attend the Grand National described in the last chapter, it went to Wimbledon and Wales that year. The Sports Editor, Peter Davies, won permission to put up an aerial above the centre court. Nobody suspected its purpose as being other than for television. But the reporter on the roof could not only describe the progress of the centre court games; he could with binoculars read the score boards of the lesser courts all round – which are very important at the beginning of the tournament. Extel's rivals were left standing – or sitting.

Gordon Dain, the new chief engineer, proved as good an ambassador in Wales. He persuaded the C.O. of a military unit alongside Cardiff Arms Park to let him put up an aerial among the military ones. From a point in this camp the excitements of the rugby football ground could be instantaneously reported.

The 1960s were a a period of rapid growth, beginning with an increase of share capital to £600,000. By July 1964 it had grown to £1,500,000. In 1960 the profits of the Company and its subsidiaries before taxation were £169,000, in 1965 £827,000, and in 1969 £1,065,000. The rest of the chapter will provide an opportunity to judge what were the main reasons which caused this great increase.

First, a few words about people. In February 1960 the man who was to lead the Company before the decade was out, Glanvill Benn, Chairman of Benn Brothers the journal publishers, became an additional Director, and in April of the following year was elected Vice-Chairman. There were several important changes on the Board. William Stevens had recently been dogged by ill health and following two major operations in 1958 retired as Managing Director in February 1959 and was succeeded by Thomas Watson. Stevens continued as non-executive Chairman for two years after this and finally retired altogether in March 1961. Watson was then appointed Chairman. Lord Sandwich retired in the following July. He gave poor health as the reason, but the fact that he had first become a Director (as George Montagu, M.P.) in 1902 might well have been sufficient. He was of more varied talents than his fellow Directors – a Trustee of the Tate, the author of a book about railways, and a poet who wrote his life story. On his resignation he was presented with a scroll, and wrote in acknowledgement:

> Thank you for the lovely Bound Scroll which the Directors and staff have been good enough to give me. I shall treasure it very much and it will be a nice legacy for my family records. I have really been deeply touched by them thinking of giving me such a beautiful present.

Six months later the charming old man died, aged eighty-eight. Francis Sheppee, his contemporary, had died in retirement the year before, at eighty-seven.

In August 1960 two senior staff promotions of men important today had been announced. Leslie Cross became General Services Manager and the contributory members' representative of the Pension Fund; and Alan Brooker, who had joined the Company the year before at the age of twenty-eight, as Assistant Accountant, became Personal Assistant to the

Managing Director. Ernest W. H. Bond, who had been Joint Secretary with Eric Martin since April, 1959, became Secretary in October, 1964.

There were three phases in Extel's development – expansion, reorganization, vitalization. And among the many factors concerned, three deserve special attention.

The first was the negotiation of a new Joint Service Agreement. Work on this had begun when the Press Association had in 1958 given the necessary three years' notice to end the old agreement of 1922, which had been twice renewed. But the fresh document had to be signed, if it was going to be signed, before the end of 1961, so the Joint Service Agreement fairly starts the decade.

Extel would have been happy to leave things as they were. But although personal relations remained as friendly as ever, the P.A. felt that their partner was getting too big a slice of the cake – or drinking too deeply from the pool. The Company, it will be remembered, had made certain reservations about what they were prepared to share, particularly in the London area. Another important point was racing. No London revenue from this went into the pool, and only a part of that earned in the provinces. The P.A. automatically supplied the provincial racing newspapers with racing news. Extel supplied the private subscribers, which for practical purposes meant the bookmakers and clubs. Up to 1960 there was not sufficient disparity in the two sources to justify argument, but the prophets among them saw the possibility of change.

Judging from George Scott's *Reporter Anonymous,* these three years were a period of mounting crisis in the P.A.'s Boardroom. In 1960 Pat Winfrey succeeded Laurence Scott as Chairman. Winfrey was determined to achieve all they wanted or nothing – in other words break up the Joint Service if not offered better terms. As George Scott puts it:

> This would mean competing with Extel in providing services to newspapers and, more important, to bookmakers, previously supplied jointly by the two agencies. This was the audacious line adopted by the P.A. in April 1960, when there was no sign of a favourable response from Extel . . . The P.A.

Board initiated research into the cost of going it alone and was ready, if necessary, to sustain a substantial investment to compete successfully with Extel.

On their side, the Company went on saying nothing much until the two sides got together round a table in the autumn. The P.A.'s negotiating team consisted of their General Manager, Edward Davies, and their solicitor, the Company's of T. F. Watson and the solicitor and Director, Lindsay Fisher. Together, coolly and without fuss, they hammered out an agreement which was considered to be fair to both sides. The division of the profits, including revenue from London and Extel provincial offices, was to be equalized by degrees over a period of years. Extel would continue to manage all the provincial centres. The new Agreement came into force on 1 January 1962. George Scott writes: 'The partnership's survival owed much to the two chief executives, Thomas Watson . . . and Edward Davies.' That he is writing from the point of view of the other side gives his words added value.

From Extel's point of view the new Agreement was a matter of give rather than take. To that extent it did not contribute to the Company's growth. But it was a firm foundation on which to build, far better than the shifting sands of competition, which prior to 1906 had proved to be costly for both sides.

The major factor by far was R. A. Butler's Betting and Gaming Bill. It cast its shadow a long way in front of it as contentious Bills have a way of doing, and the negotiations just referred to were in fact conducted under this *ombra luminosa*. But no one could forecast with any assurance what its effects would be. Even if it got past the anti-gambling lobby few people expected it to emerge half as liberal – or permissive – as it in fact did.

Betting had already been officially recognized as existing, which was not the case during much of the course of this story; but the moral point of view as expressed in law previous to Butler's Bill was interesting. On a race course you could take money from your pocket and put it on a horse. But anywhere else this was illegal – unless you had an account with a bookmaker. The moral argument (presuming betting to be sinful) appears to have been that living in sin is quite all right so long as you do it on credit, but an occasional lapse paid for in cash is wicked.

We are not, however, concerned with either morals or logic, only with the effect of the legalization of off-course betting on the Extel – P.A. racing service. This was greater than any other single factor in the whole of the Company's history. But it was not as sudden as people then concerned remember it. During the eight years from March 1961 the number of subscribers to the racing news service increased fourfold. That is what happened. The increase is tremendous, but memory telescopes the past, and the mushroom growth of betting shops was spread over that period.

This is not to say that the demands on the racing and technical telegraphic staff were not great and sudden. The shadow had become reality and the whole service had to be reorganized within the time that it takes for bookmakers and the public to react. No one could estimate this, so the preparations had to be made at once.

Leslie Cross, the General Services Manager, the newly appointed Racing Services Manager, Denis Kitchingman, Ron Witte, the Telephone Manager, and the chief engineer, Gordon Dain, put their heads together. Witte recalls that he and Kitchingman spent many week-ends in trying to work out how a greatly increased number of subscribers could be supplied by a manageable number of staff. Their main trouble was that the switchboards in use were forty years old. They appealed to the Post Office. Witte says:

> We eventually found ourselves in the presence of a Mr Troke to whom we explained our troubles. After listening to us patiently he produced from a cupboard a diagram of a new type of switchboard on which he had been working for some time, and explained that it not only improved on the multi-phone board but was also capable of easy modification for many other users. Unfortunately, he added, the official view was that there was no real demand for this type of board, and it therefore needed a firm order of a large number from a prospective customer to get production started.

The Post Office got their firm order at once. A prototype

switchboard was set up at the Cheltenham office, and worked well. The new apparatus was made exclusively for the Company until other people, seeing its usefulness, began to elbow in. But the point is that an efficient new service was in operation *before* the number of subscribers grew beyond control. By starting early and working fast the Company was in a position to exploit to the full the opportunity offered by the Betting Act.

There is an interesting sidelight to R. A. Butler's Bill. Previous to the Act there had theoretically been no off-course cash betting. Extel had given a racing service to private subscribers: what these people did with it was not the Company's concern. When betting became legal the increase in subscribers (bookmakers) was very evident in the south of England, but much less noticeable in the north. The deduction is that the north already had its betting places until the Act brought them smiling innocently into the open, with no need to join Extel's service because they already belonged to it.

The Racing service must be described, however briefly. Again, this is divisible into three – what happens on the course, at the central collecting office – which is now Manchester, not London – and thirdly the distribution from sub-offices to newspaper offices, broadcasting and television studios, betting shops and non-professional private subscribers throughout the British Isles. And there is an old adage to be used in a specialized sense: news is the most perishable commodity. Racing news begins to stink almost at once, and can attract the sharks if not put out within seconds.

The team on the race course generally consists of four – top man, floor man, commentator, telephonist. Business begins a quarter of an hour before a race. The floor man, who is in the noisy arena called Tattersalls Ring among the bookmakers, notes the betting odds and signals them by tic-tac or walkie-talkie to the top man, who is near the telephone man, who passes on these data. Meanwhile the commentator is studying the list of runners and jockeys, memorizing the jockeys' colours, preparing himself to describe the race as fast as the horses run. This he does over the 'phone to Manchester. At this central office all racing information is sorted out (there may be news coming in from several race courses simultaneously) and broadcast over the inter-office network to the provincial offices. The delay of re-transmission, which is necessary for the sake of priorities, is no more than seconds. It has been described as the echo. This echo is sent to subscribers from the local office. So if you pause at the door of a provincial betting shop you may hear the description of a metropolitan race in a local accent.

The virtually instantaneous dissemination of racing information caught the imagination of the press. The following article in *The Sporting Life* gives a doubled journalistic reaction since the writer quotes someone else; and it also shows there are private subscribers other than bookmakers:

> It was news to me that the voice of Extel, so familiar to betting shop patrons, booms out urgently with the latest news from the race track in the drawing room of Clarence House.
>
> But that was the message of a splendid piece by Margaret McCartney in Saturday's *Morning Telegraph* (Sheffield), in which she gave a fine portrait of the man behind the voice:
>
> A relaxed, dark haired, bespectacled man wearing headphones and a chest microphone, and tucked away in the corner of the racing operations room in a glass box.
>
> In the hour or so that I watched him he sat so still, so absolutely absorbed in complete concentration that I saw only his calm, clear-cut profile. Never, as far as I could see, did he so much as turn his head or lift his eyes from the collection of carefully sorted racecards on the desk before him.
>
> His facial expression never altered, only the pleasant, penetrating voice dramatized the fleeting excitements of each race as he described it.
>
> I must admit I was a bit puzzled as to how he could see so much without lifting his eyes from the racecard, but that surely is to quibble.
>
> Who, I wonder, is this man whose voice commands the attention of the Queen Mother and so many other devotees of the sport of kings?
>
> Whoever he is, he deserves the thanks of punters as his voice, clear and impartial, shivers the timbers of betting shops and tinkles the chandeliers of Clarence House.

To whom the voice belongs Extel will never say. Their

principle is that race reporting is team work, and that no one in the chain of communication is more important than anybody else. So it would be invidious to pick out any individual for mention.

Now to touch on the third factor conducive to success. At the beginning of the racing boom London and Provincial Sporting News Agency (1929) Ltd was taken over. This was an astute move, for 'The Blower' was far from being broken-winded and decrepit. It was a smallish but flourishing company which provided a distinctive personal service. If its subscribers did not get all they wanted in the routine information, they were encouraged to ring up and ask for more. It was also a clearing house for bookmakers, as is the Victoria Blower.

Extel in its quixotic way ran L. and P. as a rival to itself for a time; but not for long. It was absorbed and the provincial offices amalgamated with those of Extel, which at this time were being modernized and re-equipped.

Following the exceptionally long, cold winter of 1962-63, when there was practically no horse racing at all, afternoon greyhound racing was started. It could therefore be covered by the betting shops, as the evening meetings could not – for betting shops must close at 6.30 p.m. This meant more business for the Joint Service.

A few words about Irish racing. This anonymous record was dug out from the files:

> The Company commenced business in Dublin on 1 July 1924, at 23-24 Bachelor's Walk, which is on the north bank of the river Liffey and flows through the centre of the City. Mr John Finegan [who had been bought out by Extel] was appointed Manager and when he retired in 1950 his son Patrick took over from him. The Company moved to new offices on the south bank of the Liffey, at Fleet Street, in December 1957.
>
> Thomas Kearney was an original member of the staff. He joined the firm in 1924.
>
> Thomas Kearney: 'My job was distributing racing results which were written on tissue paper to street bookmakers and earned me the name of "Tisshy". The circuit was about seven

miles done at breakneck speed on a bicycle without ever dismounting as the subscribers were always waiting for me at street corners. Police on point duty recognized my necessity for speed and always waved me on even against other traffic, but not without first getting the name of the winner from me as I cycled past. On one occasion while attempting to overtake a Model T round a bend I finished on the bonnet and was thrown face first onto cobble stones which resulted in being detained in hospital for three days'.

It was not to be supposed that Extel gave all its attention to horse racing. That would be far from the truth. The various financial services were vitally important. There were still the Law and Parliamentary and general news. But the racing service did bring in a lot of money.

Growing richer, the Company added to itself and became a Group of Companies. London and Provincial Sporting News Agency (1929) Limited was bought in October 1962 for £300,000. In April 1965, H. G. Guest (Amplifiers) Ltd was acquired. In October of the same year the Column Printing Company Ltd (which was not trading) was used – by changing its name – as the shell in which Extel Statistical Services Ltd developed. Extel Communications branched out into the computer field. But the most exciting act of increase was achieved when Burrup, Mathieson and Company, Limited joined the growing Extel Group. That is a story in itself and as such will be covered in the next chapter.

All this was achieved under T. F. Watson's Chairmanship. But two personal decisions of his deserve special mention. The first was to present a gold watch 'to each member of the Staff who had served the Company for more than thirty years as at 30 September 1961, and to make in due course such a presentation to all employees on completion of twenty-five years of service'.

Had this been decided upon earlier it would have cost the Company a great deal of money. As it was there were a number who just missed the award by retiring or dying a little bit too soon. In 1959 Miss W. F. Speight, Confidential Secretary, retired after thirty years, Mrs D. Lamb of Statistics after thirty-four, George Smith, Chief Stock Exchange reporter, after forty-five, Percy Saltmarsh, his successor, after forty-four. And W. Ware died in retirement at ninety-three after fifty-two years' service, W. Brace after forty-eight, Thomas Bradshaw after fifty-three and William McFarlane after fifty years.

Many gold watches were none the less given in 1961. The Chairman in his capacity as Managing Director toured the provinces and made Long Service Awards in Liverpool, Manchester and Leeds. Later he went to Ireland on a similar mission. There is a splendid story of how he took his precious burden through the customs. Like good wine it has improved with age. One does not want to spoil it by being too precise. But one must make it clear that he did not try to smuggle: he declared precisely what he had. One is left, however, with the delightful picture of the Irish customs men opening a bag belonging to a respectable gentleman and finding it loaded with gold watches. They impounded it charmingly and did not release it until they had received instructions from headquarters.

The second popular benefaction for which Thomas Watson was responsible is free coffee in the morning and tea in the afternoon for the whole Extel staff. It may not have amounted to so much at the time but it now costs £2,000 a year. He might not want this known but it is told in the hope that these beverages will therefore taste still better.

In 1962 an innovation was made on the Board when an Editorial Director and a Technical Director were appointed – respectively Edward Gilling, who by this time was Editor, having succeeded Philip Burn, and Gordon Dain, whose skill had been of the greatest value in the communications revolution which resulted from the Betting Act.

For Gilling this promotion was not of long duration and had a bitter end, for in 1965 the decision was taken to close down the home news and Parliamentary services. They had been losing too much money. Thus again, as it were at a stroke of a pen, the Exchange Telegraph's purpose was drastically changed. It was inevitable, as the facts soon to be given will show: no company exists to lose money – or not for long. But it is sad that Edward Gilling should have been involved – so closely that he had to give notice to seventy colleagues. Louis Nickolls paints a

poignant picture of how he set about it when he disclosed the
Board's decision on 29 October. He called everybody concerned
to the editorial department. 'Stop all the machines,' he ordered.
The clattering ceased and there was dead silence. Then he
announced that the General Home News and Parliamentary
Services would cease on the last day of the year.

There was silence after Gilling's announcement; but not for
long. A trade union-inspired defence committee was set up.
Its first campaign bulletin, after reporting the announcement,
reads:

> Within one hour . . . strong protests were made, broad-
> cast and published.
>
> The storm grew over the weekend, with widespread com-
> ments critical of the management's decision. At 6 p.m. on
> Monday, 1 November, you decided unanimously to set up a
> Defence Committee and to embark upon a nation-wide
> campaign to prevent the proposed closure . . .

A letter was sent to George Brown, then Minister of Economic Affairs, and to Ray Gunter, Minister of Labour. T. F.
Watson was summoned by Douglas Jay, President of the Board
of Trade, to explain the situation.

A *Times* article quoted T. F. Watson as saying 'No national
newspaper pays us more than £1,575 a year for the home news
service. It is not as much as a good reporter would get'. He also
referred to the large union pay demands.

The Metropolitan Court Reporters Association sent to the
Ministers of Economic Affairs and of Labour, and to Douglas
Jay, a letter which made this point:

> In our task as Court reporters we are sometimes competi-
> tors of the Exchange Telegraph Company and sometimes
> employed by it. We have learned to respect its distinctive
> character and its reliability and we feel that the absence of the
> Exchange Telegraph service would be a serious deprivation.

It was true enough that Extel's news service was distinctive.
An agency's main purpose is to give out facts as they happen, at
once, rather than to write news articles, which is the follow-up
job of the subscribing newspapers' reporters. The agency's fact

comes first, and if it is an example of instant reporting, and therefore ahead of all others, it is called a flash.

The Exchange Telegraph had become famous for these, although there was a notorious mistake in 1941 (which was however corrected before publication). The Exchange received from a reporter a telephone flash which was taken down 'The Tiger died at dawn', and which should have been, 'The Kaiser died at Doorn'.

More recently there have been famous flashes which newspapers printed as photographic reproductions of the tape message. They were catholic in subject, as news must be. On 19 February 1960: 'FLASH – THE QUEEN: A BOY 4.2 PM XTL.' and almost simultaneously: 'FLASH – BOTH DOING WELL.' In politics, on 16 October 1964: '4.26 PM XTL FLASH – MR WILSON KISSED HANDS UPON APPOINTMENT AS PRIME MINISTER – OFFICIAL.' On 6 March 1965: '7.24 AM XTL FLASH – GOLDIE LONDON ZOO'S ES-CAPED EAGLE WAS RECAPTURED 6.15 AM.' And at the close of the case about the pornography or otherwise of *Lady Chatterley's Lover:* '2.59 PM XTL FLASH – LOVER NOT GUILTY.'

The closure of Extel's news service was raised in both Houses of Parliament. The Liberal Party put out the statement: 'The Monopolies Commission should take immediate action to ensure the continuance of a competitive service.'

It was the *Guardian* which answered this Liberal logic:

> If the P.A. finds itself alone, that is not its fault. Extel is giving up the reporting of parliamentary and general news of its own choice and for commercial reasons. The company cannot be required to continue a service which it finds un-profitable in order that some newspapers shall continue to enjoy a multiplicity of news sources.

The *Journalist* summed up in December:

> Trade Union efforts through the P.K.T.F. during the past month to save the Exchange Telegraph's Home and Parlia-mentary services ended in failure. But they were worth while if only because they clarified the issue . . .

Extel, which started in 1872 as a financial service, was never in the same position [as the P.A.]. At the end it was serving most (but not all) national dailies and Sundays, a handful of provincial subscribers (who had to transmit the service from their London offices), the B.B.C. and other overseas agencies and three foreign papers.

These last two groups paid more than the British nationals put together. In fact, an N.P.A. paper paid annually just about enough to cover the minimum rate for one reporter, and the B.B.C. decided to drop the South-Eastern cover by Extel from the end of this year.

Clearly the Home and Parliamentary services were not an economic proposition.

To make them economic would have called for many more provincial subscribers or for a five-fold increase in the income from national papers, bringing in an extra £80,000 a year.

And so at 6.30 p.m. on 31 December 1965 there came this final flash:

'EXCHANGE TELEGRAPH GENERAL HOME NEWS SERVICE IS NOW CLOSING DOWN. GOODBYE. END.'

On the same date the Company withdrew from the Joint Law service, leaving it in the hands of the Press Association.

Apart from its sadly disappointing professional end, Edward Gilling's would have been a classic success story. He was one of a family of eight, living at Lewisham. When his father died five of the children were under fourteen. Ted left school early and got an office boy job at the Admiralty. He fell down some steps and broke an arm. Thus forced into temporary idleness, he accompanied a friend of his own age who was on his way to an interview at the Exchange Telegraph. Both boys got a job, Edward Gilling as a messenger in the editorial.

He graduated to telephonist (Philip Burn gave the typical sequence in his evidence to the Royal Commission on the Press). Then he became a sports reporter, then moved to general news. This was in the days, nearly fifty years ago, when competition with the P.A. and Central News was at its height. Gilling had a way of getting his story in first. 'As a writer he may not have

R

been much of a stylist,' Nickolls says, 'but he could get to a telephone faster than most people.'

In 1929, eleven years after he had got the job of messenger,
this effervescent young man from south-east London had a most
surprising break. He was appointed the Company's Court
correspondent, accredited to Buckingham Palace.

As has been said, he received the M.V.O. He also, loving
horses and cards, enjoyed himself at Royal Ascot (in grey top
hat and morning coat), and at Crockfords. All the time he was
growing in professional stature. He had a nose for news and a
personality that got him anywhere.

In 1939 he became a war correspondent: as such his record
has been briefly given. With the return of peace he became News
Editor, then Deputy Editor. When Burn retired he became
Editor and later, Editorial Director. When the news service
went he retired after forty-eight years' service, and lived until
1970.

In May 1963 the sports ground was sold. Backing for the
Sports Club had been waning since the war. Another ground
was rented for the enthusiasts who still remained, but again one
sees a change of direction – in recreation now, not work.

Perhaps it was appropriate that Charles Wills resigned from
the Board the following year. In any case it would have had to
happen soon, for he had been in the Company for sixty-six
years, and was eighty-four. Like so many with long service
behind him he did not live long in retirement, but died suddenly on 23 October 1964.

He had started, it will be remembered, as a junior clerk aged
seventeen. Climbing slowly and steadily up the ladder, never
pushing anybody out of his way, he had become Manager, and
probably the most modest and unobtrusive Director that the
Company ever had. It is no exaggeration to say that everybody
loved him. He would help anyone with his problem and work,
give praise, thanks, and often a cheque for a good job done; and
he joined wholeheartedly in the recreations – especially cricket.
The writer has talked with old members of the staff who have
enthusiastic reminiscences about fixtures he arranged, especially
against the Acorn, a club of Wills's home suburb, Golders Green,
and also the P.A. If an Extel man made a good score in those

matches he was taken to Spaldings to choose a new cricket bat. Such things are never forgotten. Wills is remembered as 'a real sportsman.'

On the more material side he, who was active in launching the Pension Fund when its capital was £27,000, saw it reach over £1 million before the died.

In the late 1960s the Company, self-critical, called in a business efficiency organization. These bright-eyed people scouted round the office for a year: it seemed longer to those of the staff involved. Possibly something less than perfect was discovered and reported upon; but if so, the writer, not being assisted by anyone of that calibre, has been unable to discover what it was.

In 1966 the Company reorganized itself into a parent Holding Company and a number of subsidiaries. This will be the subject of the family group in the next chapter, for it continues. Here we are concerned only with tying knots at the end of strings which did not reach into the 1970s. The annunciator system at the House of Commons was terminated on 30 September 1968. Everyone concerned was entertained to a farewell dinner party by the Speaker, Dr Horace King. As soon as the Stock Exchange decided to build itself fresh quarters it meant the end of the Bartholomew House telephone switchboard and the Extel call rooms for stockbrokers. This service started in 1882 and the switchboard cords were finally cut out on 6 February 1970. This sounds sad, and it is, unless one accepts progress as a justification for everything. But lady telephonists are as resilient as they are efficient – and these ones were magically competent, pushing the rabbit into the correct burrow without any of the usual clues. Nothing but gaiety was to be detected on any face. Another loss: the linesmen's task turned finally to cutting out both underground and overhead wires, for G.P.O. lines were now used exclusively.

There were also resignations. Gordon Dain resigned on grounds of health on 31 December 1967. And next year T. F. Watson resigned as Chairman and Managing Director. Like W. C. Stevens he had, by early Exchange Telegraph standards, held the post of Chairman for a comparatively short time. But the term of office of both had been full of activity and success.

Glanvill Benn succeeded to the posts of Chairman of the parent and the principal operating companies with Alan Brooker as the Managing Director. In 1970 Brooker became both Chairman and Managing Director of the operating company with Benn remaining Chairman of the parent company.

As has been said, and as is evident, the Company has grown. One of the clearest indications of this may be read from the increasing number of shareholders. W. C. Stevens referred to their fewness in his evidence to the Royal Commission on the Press, and to the fact that shares were handed down like heirlooms from one generation to another. As late as 1955 there were only about one hundred and fifty shareholders. The number was increased when the shares were subdivided, and considerably increased again when the staff were encouraged to become shareholders, with credit assistance. Now the figure is over a thousand. W. C. Stevens must have been glad that the number was so small when he was Secretary, for Wilfred King in his punctilious way insisted that all circulars must be signed by hand. Such living memories still bind one to the old small Company.

# CHAPTER 15

# *Burrups — and the Rest of the Family*

THIS CHAPTER concerns the vitalization of the Group – new blood, new organization, new methods.

On St Valentine's Day 1952, a Boardroom minute was recorded:

> The question of acquiring a controlling interest in a suitable printing business and/or some form of light engineering activity was considered and it was agreed that preliminary inquiries should be made with a view to Board consideration of any proposition which appears to achieve the object in view.

The reason for this decision must, partly or largely, have been the Company's often discernible wish for self-sufficiency. A lot of paper was used. It was a long time since the yearly requirement of the tape machines would have encircled the world. By now the accent was on column printers – the broader paper which would have wrapped it up more thoroughly. Apart from the machines, the consumption of paper was always increasing – as it has a way of doing. In May 1948 the Board considered taking over Thames Paper Supplies and converting it into a private limited company. This was one venture where all went right from the start. By the following October everything was signed and sealed. Paper without print is like bread without butter, and three and a half years had gone by.

The acquisition of a printing business looked like being achieved quickly. By December 1952 not only had a suitable firm been discovered – Eaton and Surman – but the verbal negotiations with William Norman Kitchin, the proprietor, had been completed. A new limited company was to be formed with a nominal capital of £10,000 and William Kitchin as Managing Director. The other Directors would be General Anderson, W. C. Stevens and T. F. Watson.

But there was many a slip 'twixt Kitchin and the Company. A month later the proprietor stated that he was unable to proceed with the sale 'owing to taxation difficulties in regard to his proposed pension'. Further negotiations followed, however, and this obstacle was overcome by arranging for an annuity for Mrs Valkyrie Kitchin. The new company was to be launched with a doubled capital on 1 April. The date proved inauspicious. The husband of Valkyrie fell ill and the deal was killed.

This proved a blessing in disguise, although Extel had to wait a dozen years to find what it wanted. In 1964 it negotiated with Burrup, Mathieson and Company, Limited, which had done certain financial printing for Extel for many years.

Burrups was exactly what Extel wanted, an old-established firm with a reputation for high-quality printing in the City. Burrups on its side wanted the deal to go through. It had had a number of offers which it had turned down for one reason or another. But here was a proposition which offered the infusion of fresh life-blood and security to continue working in the same distinguished way while retaining the company's identity.

Burrup, Mathieson had some twenty shareholders, most of whom had inherited their shares as they had their Chippendale furniture, their clocks, chandeliers and cats. Samuel Bartlett, the Managing Director and Chairman, travelled round the country from Lands End to the far north getting the necessary signatures for the deal to go through. 'I was a salesman in my early days,' he says. But it was an excellent arrangement from both sides – a satisfactory marriage.

At the head of the notepaper of Burrup, Mathieson and Company, when their address was 31 Throgmorton Street, as it was until after the war, there was printed in small type:

Destroyed in the Great Fire of London 1666
Destroyed by fire in the first Royal Exchange
Destroyed by fire in the second Royal Exchange

Destroyed by fire in the attack on London, 29 December 1940

It is unusual to be proud of being burnt down. Yet one may well be proud of it when the habit started as early as the seventeenth century. To be precise, of course, it is not the fires that Burrups are proud of but the fact of having survived them, and, more than that, lived through the stresses and rivalries of City life during three hundred years.

Burrup, Mathieson are City printers with roots which reach almost to the beginning of the profession. They were established in 1628; and the parent Company, nearly two and a half centuries younger, is naturally proud of them. We will glance at the firm's long history – the main points only, for otherwise one becomes buried in ancient documents, which is not what one wants when approaching the end of a centenary volume and at last coming within sight of the present day.

Burrup, Mathieson is not a family firm in the sense that a member of the same family has been proprietor or on the Board since its inception. Had that been attempted it would probably not have survived, for such organizations tend to become enervated as if from inbreeding. New blood and with it new ideas, new drive, have constantly been brought in. And this has been done, as a glance at the family tree will show. It will be seen that, according to the old custom, the name of the firm changed with different ownership, or shared ownership.

The original Burrup's shop displayed its dolphin sign in 1730. The shop was part of the second Royal Exchange, destroyed by fire in 1838, and the amiable heraldic animal is now in the Guildhall Museum. Incidentally, the end of the second Royal Exchange was most dramatic. As the flames spread the tower began to lean, threatening to fall across Cornhill and crush the residential houses on the other side. As it tottered in the minutes before the destructive crash its bells began to sound, chiming out the tune of 'There's nae luck about the house, there's nae luck at a' ' – a remarkable understatement by a doomed building.

When the Royal Exchange was rebuilt, Blight and Burrup were accommodated at No. 12. Some years later John Burrup acquired sole control of the business. He had twin sons, John and William. They were good printers but their enduring fame derived from other interests – of which more in a minute.

William Burrup, who became increasingly eccentric with age, pressing handfuls of cigars on the friends he met in the street and holding up the City traffic like a policeman, had a son named Frederic. F. W. Burrup was a handsome, charming, and debonair young man, but a demon for business. He joined up with the printer Gerald Mathieson of No. 11 Royal Exchange, next door. Their combined forces were formidable, and in 1897 they took over Stephens, Hayter and Company – 'an amalgamation was effected' is the pleasant phrase employed in the records. This firm, under the style of Stephens and Meredith, is recorded as trading in 1628 in 'ye Stationers Arms at ye back side of ye Royal Exchange'. Thus, through one side of the family, Burrups can trace their antecedents back almost to the days when stationers occupied stations, or stalls, and certainly to the period when they printed books. For the Guild of Stationers consisted of pioneer printers in Caxton's day.

Now, reverting to the nineteenth century, a paragraph is due to the twins, John and William Burrup. One can see them in the still not overcrowded City, strutting in their tall white beaver hats. The story is still told in Burrup's office of how, about noon, they would stroll into the cashier's office and take a handful of gold sovereigns out of the till 'for lunch'. But they were no mere coxcombs, far from it. They were closely connected with the Surrey Cricket Club for half a century. John joined the club on its formation in 1845. In 1858 he became Secretary, and remained so for eight years. During this period there was a crisis concerning the club's ground, the Oval. The proprietors were letting it out for walking matches and poultry shows. The Duchy of Cornwall refused to renew the licence. They almost handed over the ground to housing developers. But John Burrup stepped in with his energy and charm and secured the transference of the management of the Oval to the Surrey Cricket Club. This achievement was obsequiously recorded to the credit of the bountiful and generous Duchy. No doubt it deserves to be so described. But without John Burrup the Oval would be the name of a group of flats or factories.

When John resigned as Secretary, William took over. He remained in office until 1872 – which means that he presided over the Surrey XI, captained by F. P. Miller, which beat all the counties and took on an all-England team. While William Burrup was Secretary, some Australian cricketers came over and tried to arrange that an English side should tour Australia. They failed: what they suggested was too expensive and too doubtful of success. Thereupon William put his personality on the scales . . . The team, captained by H. H. Stephenson, the all-rounder of his generation, arrived in Australia on Christmas Eve 1861 and made a triumphant tour. This was fifteen years before that quaint term, Test Match, was invented (what is any match if it is not a test?) and still longer before the pagan symbolism of the Ashes was inaugurated. But if countries have to play cricket for an urn of ashes a partner of a company already three times burnt was the right person to start it off.

Such was the company with which an amalgamation was effected in 1964. Apart from the cricket associations and deep roots reaching into the romantic subsoil of the City Guilds which have been mentioned, Burrups have proved themselves wide awake to the present – and the future. In the General Strike of 1926 the Directors and apprentices manned the presses so efficiently that they not only kept their own business going but printed issues for the *Daily Express* as well. When the tote was introduced they designed and printed the tickets. During the war they were given a special supply of paper and instructed to print documents and pamphlets which were air-dropped or smuggled to the Free French.

Security printing for City banks and companies has for long been their *forte*. Now they are going into Europe in a big way in conjunction with Thomas De La Rue and Company Ltd. Their thesis is that time is vitally important in getting a loan on the market. Modern communications make Amsterdam and Brussels, Frankfurt, Luxembourg, Milan, Paris and Zurich as accessible from London as from any centre on the Continent; and their combined service can supply effective co-operation with lawyers, merchant bankers, fund managers, and other financial institutions anywhere in Europe – or the world.

It was said that the Exchange Telegraph became a big

company on its seventy-fifth birthday. From about that date it began to have a family; some of them were adopted, being already of a certain age, others were brain-births. Of the former the first is the Central News. It has often been mentioned in this story but its history has never been directly told.

The Central News was a year older than the parent Company (a much less anomalous state than was the case with Burrups). It was born as a news agency, home and foreign, and the scope of its work may conveniently be conveyed by mention of its most famous scoops – or briefs as they were called in those days. It was the first with the news of the battle of Tel el Kebir and the capture of Arabi Pasha. It was twelve hours ahead of anybody with the fall of Khartoum and the killing of General Gordon. In the South African War it sent home the first news of the signing of peace at Vereeniging. It got in first with the death of Captain Scott and his companions on their return journey from the South Pole: the chief reporter, B. J. Hodson, was sent to New Zealand and cabled thousands of words on that heroic story. During its jubilee banquet in 1921 it was praised for its impartiality by leading figures of the Government and Press. And interestingly – for no famous British news agency escapes being thought by foreigners to be the voice of the Government – the French Prime Minister sent a telegram which contained the words:

> The agency will contribute by the powerful means it has at its disposal to strengthen still further the bonds of close friendship which unite the two great Allied countries. It will be a valuable auxiliary for the two Governments . . .

General Anderson in his evidence to the Royal Commission on the Press described the decline of the Central News and its purchase by the P.A. and the Company. Now only its City advertising remains, as a subsidiary of Extel.

There are many other subsidiaries. The family is now a large one. The stock and stem is The Exchange Telegraph Company (Holdings) Ltd with Glanvill Benn as Chairman, John Harvey Deputy Chairman and Alan Brooker Managing Director. The first generation includes The Exchange Telegraph Company Ltd, the operating company; Burrup, Mathieson & Co.

(Holdings) Ltd; Central News (City Advertising) Ltd, Thames Paper Supplies Ltd and City Island Property Co. Ltd which will be remembered in its original form as landlord of Extel House. It now handles all Extel's property business.

We noted the birth of Extel Communications Limited, when it faced the world with a little walkie-talkie set and a capital of £1,000. It is now the Group's computer subsidiary company.

Extel Statistics ceased to be a Department and Extel Statistical Services Limited became a subsidiary company in 1966. Its main function is still to produce card services giving the latest position and the history, for up to ten years past, of companies in various parts of the world. The largest of these is the British Company Information Service, which extends to virtually all the concerns (about 4,300) quoted on any stock exchange in the British Isles. For investment analysts and others who wish to study companies in greater depth 'Stats' produces an Auxiliary Service, giving their complete capital history and a wealth of ratios and other comparative figures. In 1969 it launched the first card service of five hundred Un-quoted Companies, to make more readily available the contents of the fuller disclosure which the 1967 Companies Act required of them. The number has risen already to 1,500 and demand from enquirers continues to extent its scope. 'Stats' is by no means insular, and also offers card services relating to Australian, European, Japanese and North American companies. The Japanese Service is particularly interesting and forward-looking, in that it is prepared in co-operation with the leading Tokyo centre of such information, Nomura Research Institute. Altogether 'Stats' is now sending out more cards than ever before. All are printed or folded to a size of eight by five inches, and the annual output exceeds 55½ million printed units.

In recent years several other branches of 'Stats's' business have been built up. It has a taxation service, which consists of two handsome annual volumes, up-dated by supplements issued during the year. One of them is the Capital Issues Book, which records all the complicated variations in the prices of stocks and shares which have occurred since Capital Gains Tax was brought in by the 1965 Finance Act, with ready-reckoning factors. The other book is the Dividend Record, which gives details of every dividend declared by British quoted companies during the Government's financial year. Both are widely used by the Inland Revenue and by the professional and business men who have to prepare tax assessments or deal in stocks and shares.

In 1970 'Stats' acquired from *The Times* one of its traditional publications, which has appeared quarterly since the last century. Entitled the *Extel Book of New Issues,* it gives a complete verbatim copy of the prospectus for every issue or placing of capital by British companies. Prospectuses being transient by nature, this is the only complete and durable record which can easily be held, and the familiar morocco binding can be seen on the shelves of leading accountants, bankers, solicitors and Stock Exchange members. For much more current use, 'Stats' prepares a series of monthly cards which give condensed particulars of new issues and placings. Another aspect of company activities is reflected in the Registrars Card Service.

The company provides a wealth of other company information, not on a regular basis but to meet some special needs. This aspect of the work includes statistics for newspapers, notably the great bulk of the material for *The Times* annual "Thousand Leading Companies".

Last but not least, it draws upon Extel's Financial News Service and other sources to supply the input data for FOCUS (Financially Oriented Computer Updating Service), produced jointly by Extel Communications Limited and Reuters.

Most of the 'second generation', subsidiaries of the first, come under the operating Company – The Exchange Telegraph Company Ltd – and Burrup, Mathieson and Co. (Holdings) Ltd. Some may be mentioned. Among the Exchange Telegraph's subsidiaries are London and Provincial Sporting News Agency Company Ltd, the activities of which have now been absorbed into the Joint Service, The Teletalk Company Ltd, which was called H. G. Guest (Amplifiers) Limited until Extel took it over in 1965. It does not, as it used to do, concentrate on selling amplifiers, but rather on renting and maintaining them for betting offices where you may hear them teletalking about races as you pass along the street. The new model is the transistorized Mark IV, a tiny thing, yet able to talk as loud as and more clearly than an operatic baritone.

The Victoria Blower Company Limited we know even better, having read Louis Nickolls's lively piece about a team in action – 'Okay Betty!' It is officially described as providing a private and confidential channel of communication by telephone between bookmakers' offices and racecourses. It works from London and Leeds, and it joined the Group in 1955.

Rollco Papers Limited was taken over by Thames Paper Supplies Limited in December, 1969, but kept its identity. It has now been fully absorbed into the operations of T.P.S. The old usefulness of supplying paper for ticker tapes and column printers is no longer sufficient. Both companies turn out rolls for use in data processing and telecommunications, and Ascom paper tape for computers.

Extel's activities and influence in the field of financial advertising and public relations – formerly represented by Central News (City Advertising) Limited and its offshoot, Fintel Limited – were widened and strengthened by the acquisition towards the end of 1971 of Dorland (City) Limited, the financial advertising agents and financial public relations consultants. Here we have another instance of the expansion and development of to-day's Group. To acquire the whole of the issued share capital of Dorland (City) Extel paid £360,000 cash from the Group's own resources and in addition discharged a sum of £37,000 owing by Dorland (City) to another company in the Barclay Group of which Dorland was a member.

As further evidence of recent development and diversification may be noted the formation of a Data Systems Division to handle computer terminal equipment. In this field of endeavour, Extel signed a contract in 1971 with Inforex Inc., of Massachusetts, to market the American company's Intelligent Key Entry System in the U.K. This agreement was linked with the acquisition by Extel of the Inforex interests from the previous U.K. distributors.

Burrup, Mathieson and Company (Holdings) Ltd has a number of subsidiaries, chief of which, of course, is their own operating company. Then there are Wightman and Company Limited, Lawrence Press Limited, The Millard Press Limited, Wodderspoon and Company Limited and, at Lewes in Sussex, Lewes Press Limited. All these concerns have been acquired by Burrups since they became associated with Extel. There is also Burrups Europe Limited. A large and complex family!

In the last few pages we have mentioned more companies than there were individuals to talk about in the early Exchange Telegraph days. Is the Company less personal as a result? Inevitably so, is the instinctive answer. But that is not quite correct. There are far more people on the staff, of course; but, at least within Extel House, they know each other quite as intimately as their predecessors did. Someone coming from outside as the writer has done, and being so warmly accepted, would be very obtuse if he failed to appreciate the chaffing intimacy which still exists. The seniors talk sentimentally after lunch of the family feeling of long ago. I believe that they take the family feeling of the present for granted. The staff is divisible into young men and men who are approaching retirement age – if they have not slipped past it without anybody noticing – into young girls and considerably older girls. But there is no communication gap between the groups.

Neither is the Board divided in the heaven-and-earth way it used to be from the staff – less so with Directors scattered thick as nuggets in Bonanza Creek now that there are so many companies to direct. The minutes, which I have seen quite enough of – being probably the only person who has read them from start to finish – are quite as much concerned with individuals as they ever were, in fact probably more so. I am the richer for many vividly remembered incidents during the last couple of years. The linesmen fooling like boys, yet working efficiently on a rooftop or under a busy street; the cheerful ladies of the telephone exchange; talks with pensioners when one saw the past shining in their eyes; old troubles now turned into good jokes; the smiling commissionaires and the eternal messenger boys; the mini skirts; the tea trolley . . .

Finally, I can't help fancying what would happen if by some magic one Monday morning the Directors and staff of the early days walked into Extel House in place of the present incumbents. Wilfred King would surely take it in his unhurried stride. Sir James Anderson would navigate the vastly greater company as surely as he did the giant *Great Eastern*. Higgins would set about improving the machines. It would work, one feels, because the spirit is the same. The Exchange Telegraph has grown a great deal in a hundred years, but it has not outgrown its roots.

# *Index*

s

Press Association (*cont.*):
rejects suggestion of joint control
of Thames Paper Supplies, 192–193;
Pat Winfrey succeeds Laurence
Scott as Chairman, 201
Price-cutting, 54
Racecourse Association: asks higher
fee for telephone facilities, 193;
agreement reached, 194
Racing news service: *The Times* report
of Henry Wilks case describes method
of transmission, 40–42; monopoly
at Brighton, 47; difficulties in
collection and transmission, 68–73;
life of racing reporter, 149; methods
of communicating odds and results
(including Alf Pepper's telescope),
149–152; use of walkie-talkie when
demands for telephone facilities
excessive, 194–197; R. A. Butler's
Bill causes reorganisation of service,
203; service described, 204–207;
*Sporting Life* on voice of Extel, 205
Radio Communication Co., Ltd, 118
Registrars Card Service, 222
Reith, John, 121
Relay offices: first office at Grimsby,
112; increased in number, 140
Rendlesham, Lord, Jockey Club
steward, 59
Reuter, Paul Julius, 10
Reuters: news service started, 50;
reports relief of Mafeking, 68;
position in agency rivalry, 74–75;
copyright in broadcast news, 120;
buys products of Thames Paper
Supplies, 193; pays more for Extel
services, 198; produces FOCUS
jointly with Extel Communications
Ltd, 222
*Reuter's Century* by Graham Storey, 115
*Revue International de l'Electricité,*
'Telegraphe Imprimeur Higgins', 99
Ridley, S. Forde, statistical depart-
ment, 110, 148
Ritchie, Mr, Alderman, 58
Roberts Statistical Company, 110
Roberts, Major Douglas, 110
Robbins, Edmund, Manager of Press
Association, 75, 76, 77, 84
Roetener, Madame, 159
Roland, R. F., sub-editor, 152, 156

Rollco Papers Limited, absorbed into
operations of Thames Paper
Supplies Limited, 223
Roneo Company, the, 109
Royal Commission on the Press,
1947–1948, Company's evidence
describes its structure, aims and
policy, 176–180
Russell, Professor John Scott, 4
Russell, W. H., 6–7
Russo-Japanese war, Company's
correspondent at, 68
Rutherford, H., chief reporter, 108
Ruttle, A. E., Berlin office, 152
Ryder, Commander R. E. D., corres-
pondent, 160

St. Nazaire raid, 159–160
Sala, *Daily Telegraph* reporter, 42
Salisbury office, 148
Saltmarsh, Percy, Chief Stock
Exchange reporter, 208
Sandown Park racecourse, 186
Sandwich, Lord (George Montagu,
M.P.): director, 85; at 64th General
Meeting, 142; in 1935, 147; retires,
receives presentation scroll, dies, 200
Satterthwaite, Edward Fowler, 13
Saville, F. W., staff member, 155
Schrimshaw, Mr, Secretary of
Albert Club, 55, 56
Scott, George, *Reporter Anonymous,* 74,
75, 76, 112, 201–202
Sculpture for new building, 182–183
Security printing, 219
Semaphore apparatus, 2
Services provided by Company: call
system by fire boxes, 22, 25–27;
New York Stock Exchange and
Paris Bourse prices, 20; call system
of American District Telegraph
Company, 20; call system in
Stock Exchange, 21; stock and
shares prices from foreign capitals,
21; stock market service to
newspapers, 22; general and racing
news, 30–33, 38–39; foreign service,
30–31; clubs receive telegrams, 31;
parliamentary information, 46;
news sheets in hotels, 46; police
instruments supplied, 46; telephone
boxes installed in Stock Exchange, 46;

*Printed in Great Britain by
Wightman & Co. Ltd., London.
A member of the Extel Group*